# The Management of Time

# The Management of Time

### JAMES T. McCAY

Prentice-Hall, Inc.

Englewood Cliffs, N.J.

Library of Congress Catalog Card Number 58-12514

Thirtieth Printing . . . . . May, 1974

PRINTED IN THE UNITED STATES OF AMERICA
B & P

# Preface

## The New Test for Leadership

*What Makes a Leader?*

Is a man a leader when he is responsible for some people?
Is he a leader when he is the most intelligent man in the group?
Is he a leader when he is the most aggressive man in the group?
The answers to these and similar questions may be either yes
or no. What factors determine the answer?

A man is a leader to the degree that:

1) he has a following

*and*

2) his following is voluntary

*and*

3) he demonstrates to people the best method of getting what they want

*and*

4) he is the best man in the use of this method.

Notice in this list of factors that the emphasis is on the future.
People see a leader as a model. His behavior shows them how
to get what they want most. They see their pay as merely the
means of providing necessities for living in the interval before
realization of their goal. They continue to give loyalty to a

leader only so long as he keeps proving without doubt that he is the superior man in the "best method."

The platoon under attack will, for example, ignore their commander if another of their number proves by his actions that he alone can pull them out of it successfully. If a business meeting has proved sufficiently frustrating, the members will quickly shift their allegiance from the chairman to the one who offers a workable way of winding up the session. The needs of the led determine the eligibility of a man for leadership.

### Trends in Leadership

As conditions in the western world have changed, the focus of people's needs have changed with them. With the emergence of new needs, people would seek out a different kind of leader to show them how to achieve their goals. Classified on the basis of kinds of leadership there have been four ages, and we are at the moment in a period of transition to a fifth. I will call these the ages of conquest, despotism, commerce, organization, and innovation.

### The Age of Conquest

In the age of conquest the odds against a man's survival were high. Staying alive was the preoccupation of most men. He who was strong, aggressive, bold, and fearless stood the best chance of living. Men sought him as their leader. He, in turn, had to prove his qualifications repeatedly through trial by battle. His leadership (and his life) usually ended with his first fight lost.

### The Age of Despotism

As men developed civilizations, they improved their chance of survival. Now their greatest threat came from conquerors.

They yearned for safety from attack. He who was powerful was the safest. Men selected the awe-inspiring ruler as their leader, and ushered in the age of despots. The model for leadership became the all-powerful master; the dominating lord who, nonetheless, guaranteed his subjects safety in return for their loyalty and taxes. His method of maintaining his power was through political action. Although he kept a standing army, he used it as only one of his pieces in a continuing game of intrigue. The techniques of political intrigue have been described by Machiavelli in *The Prince*.

### The Age of Commerce

By the time of the industrial revolution, the principal danger to public safety had become the political moves of the despots. People now shifted their support to leaders who could show them how to get a higher standard of living. These men were the newly emerged bourgeoisie of the age of commerce, the captains of business and industry who were successful, shrewd, and influential. They earned their reputations as leaders by their skill in bargaining. They speculated for high stakes. The best of them accumulated vast fortunes. Like their predecessors of other ages, they gradually increased the exploitation of their followers while giving them less and less in return.

### The Age of Organization

By the early 1920's the revolt against the leaders of the age of commerce was well under way. By now people were beginning to struggle for a feeling of belongingness. They searched for men who could show them how to fulfill these new longings. As the age of organization came into its own, the measure of leadership became the capacity to organize; the leader must be a man who could create and protect the group. New standards of conformity began to make themselves felt. It was no longer fashionable to be wealthy; no longer advisable to hold strong opinions of one's own—it was necessary to be accepted. Managers and union bosses alike, dutifully went back to school to learn the latest techniques of group decision, human relations, communications, and participation. Staff departments burgeoned with experts advising experts on the most acceptable way to gain acceptance for procedural and organizational changes. Management consultants multiplied to meet the demand for new ideas. Managers were often too busy administering their organizations and attending group meetings to create their own ideas. The goals of business became a matter of group discussion. Confusion and duplication became widespread. Meanwhile, as sales continued to rise, costs rose even faster. Managers became increasingly concerned over the steady fall of their profit margins.

### The Age of Innovation

Concurrent with the development of the age of organization was the compounding growth of scientific research and technology. Accelerated by the demands of two world wars, technological innovation reached such proportions that obsolescence became a pressing problem. The launching of Sputnik provided a dramatic fanfare for the next feature on the stage of international affairs. The age of innovation had arrived. The demands of national security guarantee that this age will have a quick and enthusiastic reception.

On the domestic scene, people for the most part have satisfied their needs to belong. They are becoming tired of the dreary monotony of the same clothes, same cars, same houses, same small talk, the same pointless meetings. The organization men are beginning to strive for personal recognition. They want to be somebody. Who will be their new leader?

## The Leaders of the Age of Innovation

The rate of innovation is accelerating. Products and methods more often become obsolete before they are through the stage of final planning. The excursion of the age of organization is threatening to become the death ride of a vehicle out of control.

The man of the hour is the one who can handle the complex problems created by the increasing speed of invention. He is not the man who is strongest, craftiest, shrewdest, or most popular. He is the man of exceptional originality. He is the man who has disciplined himself to keep acquiring new knowledge and skills. He has created new production concepts, marketing concepts, approaches to financing. If he is a top-notcher he is party to new concepts in areas like atomic energy, the conquest of space, civic planning, education, or international affairs.

In business, the transition in leadership can most readily be seen taking place in the growth industries: electronics, missiles, atomics, plastics, and the like. Those firms that have the highest quality of leadership in innovation are attracting the creative people. These come not for the money but for the promise of association with men who excel as thinkers. Soon the word gets around. "Aladdin Corporation is the outfit to get into. They've got Axelson, Sheridan, and Bronowski." Aladdin surges ahead while their competitors face the desolation of an aging product line and complaints over seniority and employee benefits.

## The New Test for Leadership

Men can be given positions. They cannot be given leadership. They take leadership by demonstrating to their followers how to get what they want most. In this age they must demonstrate originality. It is no longer enough just to be clever; to add, for example, another head light or tail light and claim a major new automotive design. The new test for leadership demands a quantum-like leap in originality, an abandonment of the old for a concept that is vastly superior. In this sense, the man with the original idea leads, regardless of seniority, position, wealth, or training.

The most striking aspect of obsolescence today is its suddenness. Regardless of size, assets, or prestige, a company can be prospering one day and faced with ruin the next. The only insurance against this threat is creative talent.

Men of originality cannot be developed en masse. Rather, they search out for themselves the sources of knowledge and skills they need for their growth. Above all, they search for a creative leader who can coach them by his example.

The message for business is clear. The companies that survive will be those that learn quickest how to discover and encourage the growth of originators. In particular, they will be companies whose managers foster enquiry and demonstrate originality in their decision-making.

The message for any man who aspires to leadership is also clear. It is no longer enough to have a college degree, to be conscientious on the job, and to work hard. To achieve and hold a position of leadership in this age of innovation a man must spend a part of every day in self-development. He will need to add continuously to his resources of knowledge and skills. To do this he cannot afford to be a spendthrift of time. He must learn to guard it with resolution, to handle it with precision, and to invest it with acumen. Then he will find he has a wealth of time for leadership and development.

# Acknowledgment

This book is in effect a record of my venture in self-management. It began in 1947 when Dr. J. S. Bois suggested that I study the writings of Count Alfred Korzybski. It led me successively to an interest in the fields of psychology, anthropology, communications, human relations, comparative religion, and philosophy. Students of Korzybski's "General Semantics" will recognize many of his ideas woven throughout this book.

Throughout the venture Dr. Bois has acted as my guide, counselor, and teacher. I have borrowed his ideas and dressed them in my own vernacular. The extent of my indebtedness will be immediately apparent when you read his recent book, *Explorations in Awareness*. Not only did he utter no word of complaint as I adopted his brain children but he read my manuscript as it developed, and offered many helpful suggestions.

This book is in no small part the result of the efforts of my editor, Eaton B. Lloyd. As I moved from revision to revision in my writing, he never failed to create an original method of getting me rolling again free of major dents and scratches.

Finally, I would like to acknowledge the tireless work of Mrs. Dorothy Lefèvre who typed the manuscript and its many revisions while managing the multitude of problems incidental to a professional office.

# Contents

## PART III

### Skills for Managing Time

## PART IV

### Plan for Development

**PART I**

# The Meaning of
# Time Pressures

Chapter **1**

# The pressure
# of time

*"No matter how hard I apply myself and re-allocate my time, I can't escape a nagging feeling that I have more work to do and more people to see than I can satisfactorily handle."*

When Barrington said this it seemed to me that he was summing up the feelings of most of the managers I had met since the end of the war.

Barrington was a man in his early forties. He had taken over the presidency of the business from the founder some six years ago. Although total sales had grown an average of 7 per cent each year of his term of office, Barrington found that it had demanded all of his attention and ingenuity to hold an adequate profit margin against increasing competition. He had asked me to come in to explore with him what he could do to stimulate the growth of his managers. He also wanted an impartial and questioning listener to cross-examine some of his thinking on current problems and long-range plans.

We were sitting in his office exchanging ideas. He leaned forward, and looked at me questioningly.

"You've met Neil Carroll?"

I nodded.

"Well, Neil has been managing our Special Products Division for the past two years. He has the whole Division humming. So much so, in fact, that I now consider Special Products has the greatest growth potential of any of our divisions. The trouble is that I can't get Neil to see that he won't be able to keep up his present rate of expansion unless he takes the time to build a competent group of men around him. Neil agrees that he has a manpower problem and claims that he's doing everything he can to meet it. He's not moving fast enough though. On top of that, I feel uneasy at the prospect of finding an adequate replacement for Neil should anything happen to him. Certainly he has no one that could take over, and I can't think of any man I could afford to take out of one of the other divisions."

As Barrington spoke, I pictured my meeting with Neil Carroll the previous afternoon. When I walked into his office he was leafing through a pile of employment application forms. He told me he had already gone through more than seventy applications in the past three months, had interviewed some fifteen men, and still hadn't uncovered anyone who could fit in as his assistant. Judging from the files on his table, it appeared that Carroll had a backlog of work hanging over him.

One particularly thick file he put away in his briefcase while I was with him. With a wry smile, he said that budgets and forecasts were part of his standard "homework." He answered at least six telephone calls during the half hour we were together. In one of them I observed him trying to explain why he couldn't serve as his industry's representative for the Red Cross Drive. In the end he gave in. Apparently he had forced the other fellow into a similar position on another charitable campaign last year. So at least Barrington wasn't alone in feeling pressed for time.

"Another thing," Barrington continued. "According to the book, I should delegate more of my work. But how can I put more of a load on men like Carroll, who already have more than they can handle? To make things more complicated, I find that with the trend towards longer holidays we are operating with three-quarters of an executive staff from June through mid-October. There appears to be a distinct tendency around May to shelve projects till we are all back in the fall. Add to this almost a month lost in the confusion of the Christmas-New Year season, and it turns out that we're trying to compress a year's work into seven months. No wonder we're short of time."

At this moment the telephone rang. It seemed that the head of the Special Names Section of the Community Chest Drive had suffered a heart attack. Barrington, as head of the Chest Campaign for this year, was being asked what should be done. He asked the caller to arrange a special meeting of his group chairmen for 5:30.

When he put down the receiver, Barrington looked at me somewhat meditatively.

"Mike Anderson, who headed up our Special Names Division of the Chest Drive, died of a heart attack last night. Mike was only forty-seven. He had a job like mine over at Ajax Mills. Last Saturday, when we were teamed up in a foursome, he looked the picture of health. It makes you stop and think."

The interruption had changed the trend of Barrington's

thoughts. We sat in silence for a few moments. Then, slowly, almost as if he were groping for a way to express an oft-repeated thought, he began speaking again:

"On the average, I put in more time on this job than anyone else in the company. Usually I'm here with the early shift and don't get away until 6:30. Evenings and weekends find me spending more time with work I bring home. My wife has more or less resigned herself to being a business widow. It's not that I don't want to spend more time with my family, but I find that home is one of the few places where I can think through a problem without a constant stream of interruptions. Not that I mind the work. In fact, I find it challenging and stimulating.

"What I find particularly frustrating is the fact that so much of my day is tied up with current operating problems. I'm convinced that it's no longer enough to meet our forecasts. Unless we think and plan actively for the next five years, and speculate on the shape of the next fifteen, we'll be like the winner of a Mississippi steamboat race—first at the finish line, with only an engine sitting on a dismantled hulk.

"If I'm not focusing on long-range plans how can I expect my men to? But we just don't seem to have enough time. The answer, I know, doesn't lie in working longer and harder. I think that's what got Mike Anderson. Frankly, I don't know what the answer is, but I'm beginning to suspect that we succeed or fail, not only as business leaders but as human beings, to the extent that we learn how to manage that strangest commodity in the world—time."

Chapter

# The challenge

# of change

*We live in an era when rapid change breeds fear, and fear too often congeals us into a rigidity which we mistake for stability.*
—LYNN WHITE, JR.

The fragment of my conversation with Barrington is a composite of hundreds of similar conversations. Almost without exception, executives have expressed in one form or another their keen and ever present awareness that they haven't time to do all that should be done. Their subordinates too, at every level of management, are seldom free from a nagging feeling that too much has been left undone. They may not admit it as fear, as Lynn White, Jr., puts it; but they are uneasy about it.

Have time pressures become an inevitable part of modern business life? Must we be crushed under the wheels of progress, like the rabbit frozen to a dead stop by the glare of a speeding car? Or shall we step boldly ahead of the movement we ourselves have set going?

## The Increasing Tempo of Business

We boast of being realistic, and we are. Then, let us face the situation. Let us become fully aware of how much the tempo of business is increasing.

Take research, for instance. In one year, 1955, American business spent more on basic research than it did during the sixty-year period, 1875-1935. And, mind you, this does not include the huge amounts of money spent on applied research on methods and procedures of manufacturing, marketing, and finance. We now have hundreds of thousands of people devoting their full time to making obsolete our present products, markets, and methods of doing business. The ratio of increase and of the resulting pressure is sixty to one.

It is not only that more new ideas are created every year. They are applied sooner. Gunpowder was invented in the twelfth century, but it was not applied in warfare until two hundred years later. Atomic fission was first observed in 1939 and applied within five years. Ratio: forty to one.

Change triggers off chain reactions throughout a business; ideas fire one another at every level of management. There are more and bigger problems than there were twenty years ago; there are more numerous and more momentous alternatives. Today's decisions may commit us for years to come. We must spend much more time than before on long-range planning. Typical of the growing clusters of big problems that face management in this age of innovation, are those confronting the major airlines today.

Since 1945 the airlines, like many other businesses, have

been struggling to keep equipment up to date while serving more customers every year. The selection of a new aircraft type takes months of careful analysis by both staff and top executives. The multi-million dollar orders, when placed, commit the carrier to an aircraft well on its way to obsolescence by the time it is delivered three or four years later. On top of that, each purchase of a new aircraft type involves new inventories of spare parts, new maintenance procedures, new training programs for pilots, flight engineers, stewards, ground crew, and maintenance men, new scheduling of routes, and often new hangers and other facilities.

These are only some of the problems associated with aircraft purchases. As the aircraft get bigger and more expensive (today's airliners come at twenty million apiece, compared with post-war price tags of around one and one-half million) the problems grow in size and number too. When we consider that most airlines have been forced to purchase new aircraft types at least every four years since 1945, it's no understatement to say, "The tempo and complexity of business operations are on the increase." We have to pack within hours or days the work that formerly took weeks or months.

### Overcoming Time Pressure

Faced with growing demands on our time, what choices do we have? First, we can pass up the challenge and step down into a task that is easier. Second, we can work longer and harder. If we have a rugged constitution we may even thrive for a while under a growing burden, but we'll probably find that fewer and fewer people will care to keep up the pace with us. We will tend to become less sensitive to their needs while becoming more demanding of their time and energy. As the pressure builds up, and long before we become aware of it, our judgment will tend to become more and more distorted by stress.

A third choice is to avoid the issue altogether. We may refuse to acknowledge that the changes taking place all around us have any significance for our own enterprise. We can think of ourselves as practical, down-to-earth business men who won't tolerate any theoretical nonsense. This is the typical freezing into a fear-bred rigidity, camouflaged under the guise of "objectivity," "certainty," "stability," or what have you. The symptoms are easy to recognize:

> We become preoccupied with the details of our work.
> We become touchy about our status.
> We become critical of others.
> We look for excuses.

There was a time when it was the fashion to fight against change. By doing so many a manager felt he was doing the right thing. Sound tradition was highly respected. Today we know that the manager who *holds on* to his job soon finds himself with empty hands.

### A Creative Approach to Time Pressures

None of these fear-dictated choices can solve the problem of time pressures. We have to take a fresh approach and cope with them in a creative manner.

A simple analogy may help us here. Let us consider the principles that control the flow of a water system. The throughput of a water system can be increased by one of two methods. We may either increase the pressure, or increase the size of the pipes and valves. To increase the flow without increasing the size means forced pressure. We have seen that the pressure of far-reaching plans, urgent decisions, and competing ideas has reached a danger point within ourselves. The alternative is clear: we must look to our personal development and find a way to *multiply our output.*

There are two ways of working on ourselves to multiply our output: the way of self-management, a way that is well within the range of anyone who wants to grow rapidly through systematic personal effort, and the inspirational, almost "miraculous" way.

There is a vast literature describing the theories and methods of the inspirational way. All theologies agree that man can be united to a greater power. They all agree, moreover, that this Supreme Power becomes a source of almost limitless energy to whoever accepts its guidance without reservations. It is when they come to explaining this phenomenon that they wind up in endless disputes. To anyone in search of a way to compound his capacity, however, a fact stands out clearly. There are countless documented records, throughout the history of all lands, of men and women who have undergone some sort of transformation. In a flash, as it were, they have suddenly made an enormous jump in personal effectiveness. Whether it be called "conversion," "salvation," "union," or "illumination," the fact remains that a spectacular jump was made. Such transformations have happened to people of all kinds, stations, conditions, and faiths.

There has been a great upsurge of interest in religious and inspirational literature in recent years. The books of Norman Vincent Peale, Emmet Fox, and Claude Bristol are best-sellers. It is quite possible that this stems from a growing awareness of the power of the inspirational way to overcome increasing time pressures.

However, in this age dominated by science, only a very small per cent of the population seems to be able to make a jump in effectiveness by means of faith alone. Perhaps their demand for an understandable working theory prevents them from undergoing transformation through the way of inspiration.

There is another way of growing in a series of jumps. It does not conflict in any respect with the ways of faith—in fact, it complements them. It is the way of man accepting without fear the challenge of change set into motion by man. It breaks through the pressure of the time barrier. Its principles and methods can be used regardless of formal education, years of experience, or work load. We shall study them together in the following chapters.

# Chapter 3

# Breaking through the time barrier

*As you refine your techniques of self-management, you may expect as a first dividend a release from the pressure of time.*

## *The Need for New Working Principles*

The manager of today is racing with time. He pushes against the time barrier as he speeds up his activities. He must break through it to meet the increasing complexity of his task.

Two methods of beating time are already well known and practiced by progressive executives. One is the method of selectivity. You recognize early the parts of your work that are obsolete and can be dispensed with: out-of-date methods, procedures, products, and facilities. You discard them without remorse. You do not preserve them by entrusting them to subordinates.

The second method, that of delegation, goes even further. You delegate those necessary parts of your job that you have already mastered. This gives you time to devote to the growing edge of your job, and it gives your men a constant challenge to grow with you.

But these methods are no longer sufficient. Faced with a mushrooming workload that you can neither cut down nor delegate, you need more than time-saving techniques. The time challenge calls for a re-examination of your working principles.

One working principle was discussed in the previous chapter; namely, to overcome time pressures you need to multiply your output. Before you can come to grips with your time problem, however, you need another new working principle. This must be a principle that will spell out HOW to multiply your output.

### A Principle of Increasing Output

In searching for a principle of increasing output, let us start with an example from the world of matter–energy control. At the physical level man has multiplied his capacity to release energy in proportions that were unbelievable a generation ago. The primitive man could pick up a 1-pound stone and throw it, say, fifty feet. His energy release was of the order of fifty foot-pounds. This can be taken as the basic energy resource of man unaided by invention.

When he fashioned the first sling, man multiplied his energy by a factor of two. He could throw a one-pound stone, say, one hundred feet. He had reached the one hundred foot-pound level.

Then came the invention of the bow and arrow. Man could

project a one-pound arrow with the force of three hundred foot-pounds. This was three times the energy of the sling throw, six times the power of man unaided.

With the development of gunpowder, man could create a twelve hundred foot-pound force from a pound of this mixture. The increase was four times the previous one, and twenty-four times the initial output possible to man.

With a chemical explosive, under certain conditions, man can wield a six thousand foot-pound force with a pound of TNT. This is five times the previous increase and 120 times the initial power.

Now man has nuclear fission and a potential energy release of the order of 10,000,000 times that of TNT. He has created a giant who has the striking power of 1,200,000,000 primitive men banded together.

Let us put these figures in tabular form:

*Energy Release Through the Ages*

| Method | Results in foot-pounds | Energy Jumps Separate | Cumulative |
|---|---|---|---|
| Stone thrown by hand | 50 | 1 | 1 |
| Stone thrown with sling | 100 | 2 | 2 |
| Bow & arrow | 300 | 3 | 6 |
| Gunpowder explosion | 1,200 | 4 | 24 |
| TNT explosion | 6,000 | 5 | 120 |
| Nuclear fission | 60,000,000,000 | 10,000,000 | 1,200,000,000 |

This progression in man's capacity to release energy can be observed at three distinct levels:

I. The *material level,* where we are dealing with chunks of material as it comes in nature or is fabricated by man. This is the stage of the stone thrown by hand, the stone thrown with a sling, the bow and arrow.

II. The *chemical* (or *molecular*) *level*, when man increases energy by combining different elements and causing them to react upon one another—the stage of gunpowder and TNT explosions.

III. The *atomic* (or *nuclear*) *level*, when man reaches down to the scientifically discovered constituents of matter and releases some of the unbounded power of the atom.

As man moved from the material to the chemical level, and from the chemical to the atomic level, he made progressively larger jumps in his controlled energy release. To make these gains in output, however, he had to make comparable jumps in the range of his knowledge and the refinement of his techniques. The operator of a catalytic cracking unit in an oil refinery, for example, needs more knowledge, must keep aware of more factors, and, in general, must be more sophisticated in his techniques than a bricklayer or a carpenter. The operator of a nuclear reactor holds post-graduate degrees, is an accomplished mathematician, yet he is a tyro in his technical environment and must therefore remain a perpetual student.

A similar broadening and refining process has been taking place in the practice of business. Until the end of the nineteenth century, for the most part business was carried on in a crude fashion. This period, which would compare to the material level of the previous example, we could call the "stage of trading." Measurement was by rule of thumb. Simple records, if any, were kept on quantities of goods on hand, sales, payroll, and the like. Business was good if you sold a lot. A business was rich if it owned a lot. There were few standards, and the law of the market place was "Let the buyer beware."

By the turn of the twentieth century this picture was changing swiftly. Industrial engineers, such as Taylor and Gilbreth, had introduced the techniques of work simplification. Jobs were no longer seen in a crude, undifferentiated fashion. They were studied in depth to discover their many elements. These elements were rearranged with ever more refinement of technique to give large increases in work output for the same, or even less input. Production per man-hour increased approximately six times in the following fifty years. Scientific management had moved the conduct of business to the molecular level of the previous energy example. We could call this period the "stage of calculated planning." To work at this stage of calculated planning, managers and workers alike have had to increase their range of knowledge and techniques.

Other examples come to mind: the fortunes in gold dust recovered from sands abandoned as "worthless" by the 49ers by modern refined techniques of placer mining. The great increases in combustion rate made possible by burning coal in a finely powdered state. The multiplication of the number of telephone messages carried by a single wire resulting from the refinements of earlier telephony. In all of these examples the same principle of increasing output can be seen. It is this: *as we learn to work at ever finer levels of subdivision; as we increase our capacity to discriminate; we can increase output in the areas under study*. I call it the principle of refinement.

### Breaking Through the Time Barrier

The trend in business practice is towards greater sensitivity of control and complexity in decision making. The increasing rate of technological innovation demands and accelerates this trend. New tools and techniques of management planning and control are being devised to cope with the growing problems.

The tools, however, demand of the user a considerable range of up-to-date knowledge and refinement of technique.

The managers of this generation face an unprecedented challenge. They are the first generation whose formal education and training threaten to become obsolete before they can fully exploit it. A symptom of this experiential obsolescence is a feeling of time pressures. The demands of the job can no longer be fully met in the available time with the old methods. Because of experiential obsolescence the manager in his forties who is no longer considered an asset to his company is becoming more commonplace. He is like the Second World War bomber pilot who can no longer qualify for air crew because his experience doesn't apply to the new jet bombers.

Productivity has been multiplied by applying the principle of refinement to the management of business. You can jump up your own output and break through the time barrier when you begin to apply it to the management of yourself. You can no longer afford to be completely preoccupied by business problems. You must, like the jet pilot, give some attention every day to the broadening of your knowledge and the refining of your techniques. As you learn more about the makeup and workings of your personal organization and its surroundings, you will become more sensitive to the elements of your experience. The remainder of this book describes how you can manage your time with increasing precision. As you apply these refinements of self-management, you may expect as a first dividend a release from the pressure of time.

# SUMMARY OF PART I

1. To achieve and hold a position of leadership in this age of innovation, a man must spend a part of every day in self-development.
2. Most managers today share a keen awareness that they haven't time to do all that should be done.
3. The tempo and complexity of business operations are on the increase.
4. Increasing time pressures call for more than a "gimmick" solution. The answer to time pressures can only be found by re-examining your basic working principles.
5. The principle of overcoming time pressures is this: TIME IS RELATED TO OUTPUT—INCREASE YOUR OUTPUT AND YOU WILL HAVE MORE TIME.
6. The principle of increasing your output is this: OUTPUT IS RELATED TO REFINEMENT OF TECHNIQUE—AS YOU REFINE YOUR TECHNIQUES OF SELF-MANAGEMENT YOU WILL INCREASE YOUR OUTPUT.
7. A feeling of being short of time is very often a symptom of accumulating obsolescence of your knowledge and skills.

**PART II**

# Overcoming
# Time Pressures

**Chapter 4**

# How to increase your output

*You can increase your output as you increase your capacity to get accurate, clear, fast impressions of what is going on around you.*

### Where Experiences Are Born

The cumulative effect of scientific research is steadily making your work more complex. To meet this challenge you need to increase your output. To the degree you fail to do so you become steadily more conscious of mounting time pressures.

The working principles discussed in Part I state that you can overcome time pressures and increase your output by refining your techniques of self-management. To do this you will need to become skilled in dealing with ever smaller elements of your experience. In the light of present knowledge, what can you take as the smallest elements of your experience that you can learn to recognize and refine?

For clues in answering this question, you can look at the recent findings of neurology and psychology. Just as nuclear physics had its beginnings a little before the turn of this century, so research on the brain and nervous system has taken place largely since 1900.

All our experience, it now appears, grows out of the activities of special kinds of cells called neurons, or nerve cells. These cells could be compared to the radio tubes in the electronic computers so much in the news today. Unlike radio tubes, however, these tiny cells can hold a constant temperature, can repair themselves, and can even grow bigger. They are found in our brain, our spinal cord, and our nerve fibres. The most complex part of our brain, the outer sheath, or cortex, contains about 15,000,000,000 of these neurons. Within this mass of interconnected cells, in ways still largely unknown to us, our experiences are born. When you stop to consider that an experience is probably a group of these nerve cells being activated together, like the lights on a theatre marquee, you can get some idea of how many combinations or experiences it might be possible to make out of fifteen billion cells. No wonder that there seems to be no limit to what we can learn.

From our eyes, ears, nose, tongue, skin, and other sensitive body surfaces, millions of tiny nerve fibres carry impulses into this central computer. From this steady stream of incoming signals, our brain manages to sort out some of these signals and organize them into a series of separate impressions or pictures. You get the effect of continuous pictures since the brain, like a motion picture projector, flashes them on the screen of our attention too fast for us to sense the gaps.

If you stop for a moment and observe yourself you will find that at any moment one impression dominates your attention.

You can shift very rapidly from looking to listening, to smelling, and then back to looking again, but each shift is made at the sacrifice of the former impression. You can only see, as we say, one thing at a time. Each of these separate impressions has been constructed by your brain and organized like a mosaic out of the bits of sensory signals in a fraction of a second. Throughout each day your brain constructs tens of thousands of these pictures. These you can take to be the smallest elements of your experience.

### Controllers of Action

In what way do the elements of our experience influence our actions? Throw your imagination into high gear and join me in a science-fiction type of analogy that will dramatize the role of these pictures in our mind. Imagine we are setting out by rocket ship to explore Venus. We blast off from Earth and after a trouble-free trip land through the heavy mists of Venus. You and I have volunteered to be the first to go overboard to explore in a special self-propelled vehicle. You are at the steering controls. The only thing you can see is what appears on a television screen that relays information from antennae outside of our vehicle. We are now alone on the surface of the planet. We switch on all our receivers and start the antennae in their scanning sequence. We watch the screen with breathless attention. It lights up but no pattern appears. There is just a regular scattering of shifting lines like this:

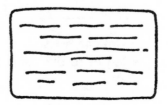

Now stop for a moment. Ask yourself this question: "So long as there is no recognizable picture on the screen, would I put the vehicle in motion?"

I think your answer would be: "I won't move until I can make out what's around me." In other words: no picture, no action.

Now let us imagine that we manage to adjust the controls of our receivers so as to get the following picture:

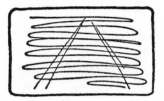

Here we can barely make out a level surface for a few hundred feet in front of us. Again, what would you do? I think you might be willing to move the vehicle ahead very cautiously to explore up to the edge of our foggy picture. But you act with caution. You move slowly and are prepared at a moment's notice to retreat.

A little later we are able to get the following very clear picture on the screen:

With a clear picture you can act with confidence. You can turn the vehicle to the left or to the right, and explore along the edge of the cliff, or back away from it. It's conceivable that our receivers might have been out of adjustment and could have given us instead of the pictures above an equally clear but false picture. In this example, if the false picture had shown a plain instead of a canyon ahead of us you would have been equally confident about going full speed ahead!

This analogy underscores the fact that the pictures in your mind control your actions. If you have no picture; if, as you say, you can't make out what's going on, you don't act. If your pictures are cloudy and confused, you act hesitantly. If your pictures are clear and accurate, you act definitely and effectively. Furthermore, the speed at which you make your pictures governs the swiftness of your action. Remember, you can't act without a picture.

### *How to Increase Your Output*

The general method of increasing your output can now be stated. You can increase your output as you increase your capacity to get clear, accurate, fast impressions of what is going on around you. The specific method of making faster and better pictures will be described in detail in Chapters 5 to 11. It involves three phases which you can work on simultaneously. They are:

1. Increasing your alertness
2. Increasing your available energy
3. Increasing your knowledge and range of experience

Here are some examples to illustrate how these factors can influence performance:

First, the factor of alertness. You begin the day with high hopes of making a sizable dent in your backlog of work.

You just get nicely into your morning's mail when the telephone rings. It's Barnes, in Production, asking what he's supposed to do with the truckload of damaged goods that has just been dumped on him unexpectedly. You tell him you'll give him an answer as soon as possible, and immediately call Smedly from Sales. Only last week you spent an hour with Smedly agreeing on what policy was to be followed on this shipment. When Smedly comes in he is full of explanations. He claims he knew exactly what to do and did it. On checking, you find that he had a clear picture, all right, but the wrong one. In any event, over an hour of your morning, plus all the other time you spent on the matter, has been invested to yield an unsatisfactory situation that promises still further claims on your attention.

What went wrong here? Thinking back you recall that Smedly seemed quite preoccupied last week when you talked with him about the shipment. Several times you had to repeat ideas when you noticed a glazed look in his eyes that showed he was off the business at hand. Come to think of it, that day you were pretty preoccupied yourself. Just before the meeting with Smedly you got a trouble call from home. Worry over that problem kept intruding into your thoughts while you were talking with Smedly. Perhaps if you'd been more alert you would have done a better job of giving instructions.

Preoccupation, or lack of alertness, is probably the greatest single factor that reduces your output. When you are preoccupied your attention is on something other than the happenings around you. You are therefore unable to get any impression of what is going on during these moments. In Chapters 5 and 6 you will see how preoccupation can capture your attention, what happens to your time with loss of alertness, and finally, how you can gain more moments of alertness and your freedom to manage your time.

Here is another example to illustrate the factor of energy and the role it plays in your efforts to make clear, accurate pictures of what is going on. You have a 100-page report on a new process. You spent the better part of the weekend going over it. This morning you are studying it once again. On the surface, its recommendations seem quite straightforward and logical. Still, you are having great difficulty in gaining a picture of the report as a whole. You are expected to give an analysis of the report to the management committee this afternoon, and the time is running out. You feel exhausted. You have a slow, throbbing headache that seems to start from a point just behind your sinuses. You press your fingers gradually up your temples in an effort to relieve the pain. What does this report mean anyway? If only your mind were clear! Your trouble here is one of being low on energy. When your energy is low you find it difficult to concentrate, difficult to get even simple ideas in focus. In Chapters 7, 8, and 9, you will find a description of the factors that rob you of much of your energy. You will also see how you can shield yourself from energy losses.

The third example illustrates the relationship between the factor of knowledge and experience and a man's capacity to make clear, accurate pictures. Your company has decided to install a system of standard costing. A team of consultants has made a complete study of the plant and has worked out new procedures, standards, and report forms. All of the department heads have been briefed on the installation and appear to be in support of the plans. After several months you are asked to check on one department that is operating with a sizable negative variance from standard. You talk with the head of the department and find that he seems to be unaware of his poor performance. You show him the cost statements and point out the figures in the red. He is astonished. He says, "So that's what those reports mean. I haven't been able to make head nor tail of them. I've been sticking them away in my drawer because they didn't seem to be of any use." On further checking, you find that the man has had practically no experience with cost statements. In fact, he has made a practice of avoiding them whenever possible. In the briefing session on the new system, he kept quiet because he didn't want to appear ignorant.

This man could not make a picture from the new control forms because he lacked the necessary knowledge and experience. In Chapters 10 and 11 you will see how you, like everyone else, are blind to certain kinds of happenings through lack of knowledge and experience. You will see how you can systematically broaden your background of knowledge both on the job and off it.

Chapter 5

# Freedom to manage your time

*You can manage your time only in those moments when you are alert to what is going on within and around you.*

### Making a Clear Picture

Before you can recognize and remember something you must first make a clear picture of it. Think back to the last time you deliberately set out to remember something, perhaps something you read. To remember it distinctly you had to give your full attention to it. In the same way, to get a clear impression of an odor, a sound, a texture, or taste, you must focus your full attention on smelling, hearing, touching, or tasting.

This point is crucial to the rest of this book. Stop now, and try the following experiment to give yourself a real feeling for this idea. Start thinking about what you were doing last night—not in general, but in detail. Now continue reviewing those memories while directing your gaze to some object nearby. Without losing your train of thought, see if you can make a clear picture of the object. Try this several times with different subjects to think about and different objects for observation. What do you notice? You probably find that you get into a sort of mental tug-of-war. Torn between remembering and observing you end up with a scrambled picture. To gain a clear picture of the object you must give it your undivided attention. How often throughout the day do you give your full attention in this way? Put another way, how often each day is your attention taken up by memories of things past or speculations on things to come? Answer this question and you will know how much freedom you now have to manage your own time. How can you extend this freedom, continue to improve your picture making, and thereby increase your output? The answer lies in an understanding of the role your habits play in your life.

### The Role of Habits

Once you have learned to do something you can delegate it to the automatic control of habit. You can shave, dress, eat, drive to the office, and do the routine parts of your work as a matter of habit. What does this mean? It means that your attention can be in other times or places than that of the task at hand. Your body carries on in a robot-like fashion as long as nothing unexpected crops up. With your thoughts elsewhere, planning, judging, remembering, speculating, or daydreaming, you, of course, are not alert to what is going on around you. So long as you are in a routine, relatively unchanging, and familiar set of circumstances you can operate reasonably well on automatic control. If the situation calls for a conscious choice among various alternatives, you must take over in full alertness and keep a clear picture of what is going on. You can't be "lost in thought." Would you like to steer a car on an unfamiliar mountain highway, for example, while looking at a program on a portable TV set on the seat beside you?

Your goal is to multiply your output. You can do this by increasing the number of clear and accurate pictures you make each day. This in turn calls for more moments when you are alert and giving your full attention to what is going on. *Any moment you are preoccupied, acting habitually, is a moment you are not free to manage your time.*

This is not to say that it is necessary or even desirable to aim for a constant state of "bug-eyed" watchfulness. Your habits play a vital role in relieving you from the monotony of simple repetitive tasks. They take over the control of your activities when you wish to turn your attention to planning, analyzing, a friendly chat, or just plain loafing. If not guarded against, however, these versatile servants can stealthily take over the bulk of your controls and impose a well concealed tyranny upon you. Here, for example, is a typical instance of how they can take control.

### The Loss of Freedom

Ned Sanders was fresh out of engineering school. He had been recruited in his final year by Goshen Iron Works, and had been eagerly looking forward to starting with them. Ned graduated with average grades but had shown a good deal of initiative both in his summer jobs and every school year. After a week's processing through Goshen's new indoctrination procedure, Ned had been planted in Production Control. The Vice President, Engineering, had given him a hearty handshake as he told him of the assignment and said, "You'll get a fine chance to see how this outfit ticks from the vantage point of Production Control." Ned went away with a determination to really show everyone what he could do. Within a week he had a dozen ideas on improvements for the department. When he showed them to his supervisor, a soured old-timer who constantly wore a harried look on his face, Ned got a curt brushoff. Ned kept working conscientiously but began giving more of his attention to thinking about his spare time activities. After all, he was finally free of the drudgery of study, and had both money and time to spend.

It wasn't long before he got married, and the following year the first baby arrived. These were exciting years, with plenty to keep Ned on his toes and interested in the goings-on around him. True, he seemed to be stuck in a rut in Production Control; he'd mastered the work in the first six months, but things had been slow in the shop and he figured when business picked up he'd get a transfer. Meanwhile he had his hands full landscaping the garden, walking the baby, bowling two nights a week, keeping up with the steady flow of trade magazines that came over his desk, and reading *Time, Reader's Digest, Saturday Evening Post,* and the daily *Journal.*

At the end of the second year Ned got a transfer to the Engineering office. He was put to work on the design board, and regained a bit of his former enthusiasm from the challenge of his new job. Along with the transfer he got a good raise. With this extra money Ned and his wife figured they could now buy a new house. Once in the new neighborhood, they felt somewhat ashamed of their old car so they financed a smart, hardtop convertible, too. The arrival of another boy ten months after the move really stretched their budget. After this, somehow, Ned and his wife never quite escaped the feeling of being on the verge of a financial chasm.

There were no further moves during the following five years, but every Christmas Ned got a small bonus and a routine raise. He came to depend on these year-end "cash crops" to bail him out of his regular December overdraft at the bank. By now he had a healthy stake in Goshen's employee benefit plan, and he no longer indulged himself by muttering about opportunities elsewhere when things got particularly frustrating. These occasions seemed to occur more often lately. Ned knew the ropes backwards by now, however, so he could turn his attention away from work with practiced skill and think of more pleasant things; the bowling tournament, the new patio he was working on, the speech he gave last week to the boys at the Spoke Club.

Then Goshen launched a big expansion program. A whole line of new, complicated equipment was taken on. Ned found himself faced with a host of unfamiliar problems. He had to work furiously to keep up with the pace. He found that much of his reserve of knowledge from college days no longer fitted the new challenges before him. Though he worked hard he seemed forever to be behind. There just weren't enough hours in the day. On top of that, when he got home at night he found it almost impossible to concentrate on the technical literature he'd brought with him to bone up on. Then the blow fell. A younger engineer was given the appointment to Assistant Chief Engineer, a post Ned had counted on. What had gone wrong?

Ned had lost control of his time. He was unaware of this because it happened so gradually. Instead, he blamed it on the company, and, in particular, he felt bitter towards the Chief

Engineer whom he now "knew" played favorites. In part he was right but he had himself to blame to at least an equal extent. He began losing his freedom when he relaxed from the discipline imposed on him by his formal schooling. There was no interested supervisor to coach him in the difficult transition from externally imposed discipline to the self-discipline of personal management and growth. He continued to lose more of his freedom as he progressively moved deeper into debt. Worries over meeting his bills claimed an ever-greater part of his attention. Another chunk of freedom went by the board when he gave himself over solely to editorialized news reports and to pre-digested articles. He seldom read a book for information, and then only those that were popular. He lost the capacity to think for himself, though he was always ready with a snappy cliché or the current headline. He exchanged his freedom, too, when he indulged in the luxury of blaming others for his difficulties rather than undergo the catharsis of self-examination.

Ned lost his freedom and the control of his time while becoming dependent. He became dependent on his work for his continuing education. He became dependent on his company for a yearly raise and for his future security. He became dependent on his neighbors for his sense of self-esteem. He became dependent on his magazines and news sources for his stock of information. He became dependent on the good will of his boss for continuing promotions. He became dependent and he became ever more hurried, anxious, frustrated, and short of time.

### Freedom to Manage Your Time

You want time to get out from under your growing burden

of problems. You want more time to explore those books other people keep talking about. You want more time to improve your golf game; to become a skilled platform speaker; to take up the piano again. More time to enjoy your children, to visit your customers or suppliers, to travel.

To get this time you must regain a beachhead on that continent of yourself that preoccupation has already usurped. You know that it is when you are preoccupied, acting habitually, that you fail to see that your listener is confused, lacks a clear picture of what you want him to do. You know that it is when you are preoccupied that you may miss a vital part of what the customer or your boss is trying to tell you. You know that it is when you are preoccupied that you speak out of turn and sometimes destroy a relationship you have spent months of your time striving to create. It is in these moments of preoccupation that the incidents are born that can swallow up hours and days of your time to no avail.

You can steadily increase your freedom to manage your time as you win back your attention from preoccupation. How you can do this and enjoy greater independence is the subject of the next chapter.

**Chapter 6**

# Increase your

# alertness

*You can increase your alertness through changing routines, through a daily period of practice, and through cultivating interests centering on observation.*

## The Power of Preoccupation

To have more freedom to manage your time you need more moments of alertness each day. Without alertness you will be unable to improve the quality of your picture making and increase your output. Without alertness you will become involved in time consuming misunderstandings, errors, and arguments. To gain more moments of alertness you have to win back some of that continent of yourself taken over by preoccupation. Before launching on this operation you would do well to appraise the power of the enemy you are about to attack.

First of all, how widespread is preoccupation? To answer this look at the people around you. Watch the people passing on the street, people eating near you at lunch time, people sitting around the room in the next meeting you attend. Look at their eyes. Is their gaze fixed or shifting? An almost sure clue to preoccupation is a fixation in the eyes. Only the most rigorous training can give a person the capacity to fix his gaze while keeping fully alert. You can safely assume that few of the people you will be watching will have trained themselves in this skill.

Fixation, which is a form of repetition in behavior, can be taken as a general clue to preoccupation. For example, people are usually preoccupied when they talk in a monotone; when they chew their food steadily without change of pace; when they hold any position without moving; when they jiggle their foot, rub their hands, doodle, or otherwise act in an unchanging pattern.

Look at the people around you for these and similar signs of preoccupation, and you will be amazed. Almost everyone will seem to be preoccupied most of the time. You will feel, if you try this, that you are surrounded by sleepwalkers. You will hear people talking but you will quickly recognize that most of the time they are not conscious of you, they will be "lost" in what they are saying. You will see people hurrying past you in the office, oblivious to the happenings around them. You will watch men in meetings making points that are unheard, voicing objections to issues that have never been raised, agreeing to ideas that have never been presented.

Now consider the forces in support of preoccupation. Remember that repetition is the basis of habit formation. When you repeat something often enough, you learn it. When you have learned it you can turn your attention elsewhere and let your robot controls of habit take over. The things that you do repeatedly or experience every day, then, are forces in support of preoccupation. What are some of these?

Look first at your network of routines. You have a routine for washing, dressing, breakfast, going to work, and many routines within your work itself. Possibly you have routines for lunch, for reading your newspaper and magazines, writing letters, going home. Then there are your routines for entertaining, the club, church, mowing the lawn, and so on. This is not to imply for a moment that you shouldn't have routines. It does mean to underscore, however, that these are all occasions when preoccupation can win you over.

Another force in support of preoccupation is advertising. You know that a keystone of advertising is repetition. Every jingle, slogan, trademark, singing commercial, and product theme batters away against your alertness. The goal of most ads is to make you want something. To the degree they succeed in creating this desire they take your attention away from the present (the state of alertness) and project you into a preoccupation with the future.

Yet another group of forces are the newspapers, picture magazines, slick magazines, and other printed material which comes in standardized makeup. Once you get used to your paper and magazines it is possible to go through them from end to end without remembering a thing. This again does not mean to say that they cannot be read with alertness. It is just a reminder that these information sources can put you in a state of preoccupation.

Finally, and, as you know, I am by no means at the end of the list, there is the endless variety of entertainments designed to distract you. Spectator sports, motion pictures, television, radio, to mention a few, all vie for your attention. If engaged in with alertness they can be a growth experience. Too often they tempt you to become preoccupied. Consider all these things and you will have a profound respect for the power of the preoccupation you plan to attack.

### Taking a Beachhead of Alertness

The attack on preoccupation, as you can see, is not one to be entered into casually. To stand any chance of gaining a beachhead you have to throw your full resources into the attack. There are three major weapons you can use to beat back preoccupation. They are: the changing of routines, a daily period of practice, and the cultivation of interests centered around observation.

First as to routines: whenever you change one of your routines you are forced to be alert in order to follow the new method. In the beginning of your struggle against preoccupation change as many of your routines as your nervous system can stand. I say as many as your nervous system can stand because changing a routine puts you under increased stress. In the same manner that you would go into a program of physical exercise in gradual steps, change your routines gradually, adding a few more each week. You can try, for example, dressing in different sequences, taking different routes to work, eating in different places at lunch time, reading informative books in the way outlined in Chapter 18.

There is a wide range of opportunities for beating preoccupation on the job. Here, for example, is what you can do in three situations that crop up in almost every business day:

1. *Meetings*
   a) Speak slowly and in a modulated voice.

b) Wait a few seconds after someone has spoken before you reply or raise another issue.

c) If you disagree with something someone has said, check yourself before voicing your disagreement.

d) Avoid any criticism, either implied or direct.

e) Avoid making excuses; give the facts if asked for them.

2. *Conversations*

a) Follow points "a" to "e" given for meetings.

b) Work on the assumption that you misunderstand what is said, and keep checking what you hear with questions calling for clarification.

c) Keep watching the speaker to see if his facial expressions and body sets correspond with what he is saying.

3. *Walking Through the Plant or Office*

a) Keep looking at things through the eyes of a buyer. Imagine you are given the option of buying anything in the company. What would you pay for as is? What would you pass by?

b) Look for methods, facilities, and products that are becoming obsolete. Ask yourself these questions: "What would I eliminate?" "What would I replace?" "What would I replace it with?"

The second way to attack preoccupation is to select some activity you can give your full attention to practicing for some time every day. Preferably it should not involve talking as this will tend to lead you back to preoccupation. Preferably, too, it should be something you haven't tried before. Among the possibilities you can choose from are: playing a musical instrument, painting, sketching, dancing, sculpting, singing. You don't have to be a stickler for holding to the same time every day for practice nor for keeping to the same amount of time every day. What is very important, however, is to practice *every day,* whether you feel like it or not. This period of practice will become the strongest part of your beachhead once you strike out into enemy territory. While practicing, watch the following points:

1. Follow whatever routine you decide upon to the letter. Change it from time to time to prevent it from becoming habitual, but don't deviate from any pattern once you've decided on it for the day. This will help you to hold your attention to the business at hand.
2. Keep your attention away from extraneous thoughts and distractions. In the beginning you will find it easier if you can find a place to practice that is free from distractions.
3. Keep your practice simple at first in order to avoid becoming preoccupied with the goal of achieving mastery of the activity. Your goal is to keep your attention on what you are doing from moment to moment.
4. Keep checking regularly, all over your body, for signs of developing tension and see if you can let them go before they build up. The skill of being able to relax while in action is particularly valuable for self-management. If tensions do build up, stop and let them die down for a while.

Soon you will come to look forward to your daily period of practice. You will find that it gives you a growing feeling of being in command of yourself. Other people will sense your growing "power to do" as well.

Third, cultivate at least one interest centering around observation. Take some time on clear evenings, for example, to go outdoors and familiarize yourself with the constellations. This may lead you to buy a telescope. Then you can explore the surface of the moon, hunt for galaxies, or study the variable stars. Buy a pair of 7 x 50 binoculars, get a copy of Roger Tory Peterson's "Field Guide to the Birds," and see how many birds you can identify. You can't study the characteristics of a star or a bird and be preoccupied too. Perhaps photography will be more to your taste. Not just snapshots, but photographic studies of the happenings around you. Or try microscopy, gardening, or prospecting—anything that demands your full attention and interest.

Throughout the day, also, as you drive your car, walk through your office, sit in meetings, interview people, cultivate the habit of shifting your gaze. You know that fixing on the eyes of a hypnotist, a sparkling gem, or a candle flame can move many people into hypnotic trance. Learn to flicker your attention over the many facets of what is going on about you, and you'll see more, hear more, and remember more.

All these new activities will demand a lot of your energy. This will be particularly so if you are presently living a fairly routine life. Unless you can learn some way to maintain large reserves of energy your attack on preoccupation may fizzle out. The next three chapters will tell you how to conserve your energy. With more available energy it is easier to keep alert and make clear, accurate pictures. With better pictures you can increase your output and overcome time pressures.

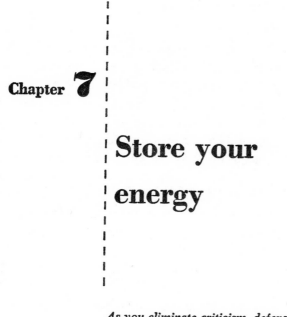

Chapter 7

# Store your energy

*As you eliminate criticism, defensiveness, and other negative factors, you will build up a surplus of energy for the better management of your time.*

### The Influence of Stress

When studying performance, it is instructive to watch what happens under conditions of high stress. By including the extremes of the range of operating conditions we can discover much that is usually hidden under normal circumstances. Here, for example, is a description of a meeting in which men are under high stress.

The meeting started at 1:30 in the afternoon and it is now 4:45. Five men are in the room, their coats and ties off, their shirts clinging to their backs from the heat. Bands of blue smoke hover in layers over the conference table. One man stares tiredly out of the window. Another is just completing his second sheet of a very intricate lattice pattern. He is completely absorbed in this self-appointed, irrelevant, dream-like drawing. A battle is being waged among the remaining three: a methods engineer, a Division Manager, and his assistant. The methods engineer has spent the first two hours of the meeting going over a division-wide revision of systems and procedures, step by step. He had been asked by the Division Manager to do the study after sufficient pressure had been applied by Head Office. The methods engineer and his assistant (the fellow who is staring out of the window) have put in three intensive months of work on this plan and two weeks in preparing their report.

Throughout the presentation the Division Manager has made only a pretence of interest. Well before the finish, the methods engineer found himself telling the story in a defensive way, a way that became almost defiant as he grew more aware of the hostility implied by his listener's behavior. At the point we look in on this meeting he has been defending his plan under open attack for over an hour. He's reached the point where he's barely able to choke down his anger when replying to a fresh criticism. The Division Manager has clearly shown that he never intended to accept any plan put forward by the methods department. The only reason the meeting still drags on is that by now no one can think of a face-saving method of escape from the situation.

Could we have recorded the whole meeting we would have found that as tension increased, new ideas in the form of con-structive suggestions and workable compromises became fewer and fewer. There is more repetition of criticisms and defenses. The last part of the meeting trails off into painful silences broken occasionally by senseless interjections.

Fatigue has set in. Don't ask the members how they feel—it is too apparent. They are dog tired, utterly weary. If they were to report on their condition they would make remarks like: "I've a terrible headache," "I've a stiff neck—feel really beat." At this point they are drained of all energy.

When our energy is low we are not able to create fresh mental pictures. It's too difficult to concentrate. We drift off; idle, irrelevant pictures form, or the screen of our attention just goes blank. Since we act on the basis of our pictures, we don't feel like doing anything when we're tired. If we want to overcome preoccupation, we must keep our energy level as high as possible.

### The Role of Negative Factors

Many factors were at work to lower the energy level of these committee members. One particular set of factors could probably account for most of the energy-leaks. We may call them "Negative Factors." They are a cluster of attitudes of defensiveness, criticism, resentment, suspicion, fear, worry, and the like. I say a cluster of attitudes because usually they are found huddled closely together like aphids. Like aphids, too, if one develops you soon have a thriving colony. They breed and suck up energy at a breathtaking pace. The relationship between available energy and negative factors is something like that shown on the chart on page 50.

Notice how swiftly the energy drops as the negative factors increase.

Two species of negative factors seem to infest most business days; they are *defensiveness* and *criticism*. They're often difficult to detect since they wear many guises. Here are some of the more common ones:

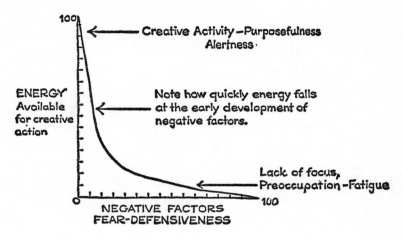

*Defensiveness*

Defensiveness often appears as an explanation. Someone asks us: "Did you send that notice out to Bob?" Immediately, we launch into a long explanation about how busy we've been; the sudden problem that came up; the people that came in unexpectedly. Our questioner, however, usually isn't interested in listening to our explanations—he just wants to know if the notice was sent and, if not, when it will be sent. We could have saved our time, our loss of energy that grows with our increasing feelings of inadequacy as we string out our explanation, and the time of the other fellow if we had just replied: "No, I haven't, but I'll get it out right now."

We go into a meeting without having made the necessary preparation. Our first remarks may run like this: "I told my secretary to line up all my files on this subject. Got the whole thing organized myself, then J.B. dropped in for a chat and the whole afternoon was shot. I took all the files home last night, and just got nicely started on them when Tommy came in screaming (you know Tommy, the little fellow who came toddling out on the lawn at our last barbecue party)."

At this point everyone shows concern, asks questions, commiserates, and otherwise gets well away from the starting point of the meeting. The explanation runs on and on, with interested and helpful advice thrown in from the sidelines. The cost of overcoming your defensiveness might well run into a half hour's time of the seven men who came in to discuss an operating problem. You could have kept quiet and used the time to brief yourself on the problem, or, if you were chairing the meeting, could have postponed it until you had made the necessary preparation.

Defensiveness frequently takes the form of rationalization as we talk with ourselves. We have done something we wish we hadn't done, failed to look after a job we had promised to do, promised to do something that now we don't want to do. We begin to figure out why it couldn't have happened any other way, or why it really shouldn't be done. We seek a justification, and, since it's almost always weak, we have to keep bolstering it up, repairing, and protecting it. This becomes an area in which we are very touchy. We can easily become resentful if anyone brings it up. We've another hole in our boat, and we've only patched it up with adhesive tape. How simple it would have been, how much energy could have been saved, if we had acted differently! If we had just made a mental note not to do again what we wished we had not done. If we had done the job we failed to do, or, if that was not possible, had apologized, without manufacturing explanations. If we had done the thing we did not want to do, or had gone to the people involved and said we didn't want to do it—again without any manufactured explanation but with a simple statement of fact. There is a

Confucian saying worth remembering: "He who justifies does not convince."

## Criticism

Criticism often appears in the guise of information-seeking. We ask a question, presumably to get information, but really to point up a weakness in the other person's ideas. For example, questions such as "How many people do you really think would buy it?" "Have you looked into all the background factors?" "Do you really think this idea is good for the company as a whole?" Next time you hear questions like these being asked, watch for the reaction of the man answering. He will probably go on the defensive. Then watch the negative factors pile up!

Questions can be asked in such a way as to indicate that the inquirer is honestly seeking understanding. Questions beginning with phrases such as "I'm not quite clear what you mean by . . ." or, "Could you help me get a picture of . . ." It's not just the words used, as you know, it's the attitude you take as you use them.

Criticism also appears often as purported statements of fact. "That idea really isn't relevant," "Your work is untidy," "His department is poorly organized," and so on. Each of these statements, you will notice, contains an adjective or adverb. They are judgments. They express how the speaker *feels* about what is going on, but they appear to be statements of fact about the event.

Whenever we make statements such as these, we almost always put the other person on the defensive. As we have observed, we also cut down our capacity to see what is going on. We can reduce the negative factors by rephrasing these statements (if they must be made): "I'm not quite clear on how your idea fits into the picture," "I don't feel happy about sending out a letter with erasures in it—would you please retype it," "I can think of other ways he might organize his department that might give better results."

Another form of criticism that can drain away a good part of our energy is self-criticism. By self-criticism I mean the various ways we condemn ourselves. If we want to keep growing, it is necessary to keep observing our actions; to judge them, and decide which we want to repeat. It is not necessary to dwell on mistakes, however. Every moment spent brooding over a past act is a moment lost to creative action. Rather than brood, the thing to do is answer the question: "What can I do now that will help most?" Once you have the answer, stop questioning and act.

To go back to the meeting—the smoke, the heat, and the hour of day were not the only factors controlling the level of energy of those present. We can all remember meetings we've attended under these kinds of conditions, yet they were meetings that left us charged with energy, "raring to go." The major factors that stole away the energy in this meeting were the negative ones: the prevailing attitudes of criticism, defensiveness, resentment, and suspicion. They could have been largely avoided.

## *Store Your Energy*

All men who have achieved mastery of the self consider the storing of their energy as fundamental. There is little point, they say, in learning skills and techniques if one hasn't the energy and alertness to use them at will. They agree that most surplus energy can be gained by cutting down on negative factors. They point out, however, that the bulk of negative factors is usually well hidden from view, that it takes a saint to be keenly and momentarily conscious of his negative factors.

The discovery and elimination of negative factors is an art, and, just as in any other art, it takes time and practice to become skilled. In the following two chapters I will describe in detail some of the techniques you can use to discover sources of negative factors, and to shield yourself from them.

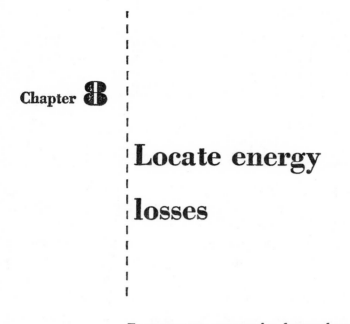

Chapter **8**

# Locate energy
# losses

*To save energy, you need to know where
you're losing it.*

### The Grid Principle

Before you can store your energy you must learn where
you're losing it. There is a well-known and time-tested principle you can use to discover energy losses that lie within your
control. It is the principle of the grid.

If you were asked to sketch a landscape and represent its
features as closely as possible, you could use the grid principle
in the following way. You could take a pane of glass and paint
on it in fine black lines a grid of, say, five squares horizon-
tally and vertically (see diagram on page 56).

You could then draw a similar grid on your paper, set up the
pane of glass between you and the scene, and fill in each square

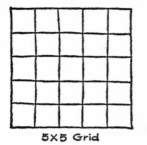

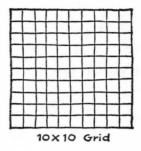

5×5 Grid                          10×10 Grid

of your paper with what you see through the corresponding square in the glass. If you made a finer grid, say ten squares both ways, you could make a big improvement in the quality of your picture. Since now there would be one hundred squares in your grid rather than twenty-five, the improvement you could expect would be more than double. Note, however, that the amount of work to make the picture would also be greater. For any task of observation, the trick is to pick a grid that will give you the optimum balance between the information received and the effort expended. The choice will depend on your purposes.

The grid principle is used in countless ways: to divide up a search area to locate a missing plane; to divide up a market to observe sales results more accurately; to divide up expenses into cost centers to see more clearly where the money is being spent; to divide up a work method or procedure to see better how it can be improved.

Most of us use a very simple grid for observing ourselves and others. I call it the body-mind grid. We sort out our observations into two groups with this grid. We observe a person's physical condition—his body—and we observe his mental condition—his mind. We say, "You look kind of ragged this morning," "You're gaining weight, aren't you?" "He's in perfect condition." We also comment on the so-called "mental" aspects: "He's not very bright," "I feel really sharp today," "Joe seems to be very depressed these days." This simple body-mind grid is all right as far as it goes, but we can observe much more if we use a finer grid. You want a grid for discovering energy losses in time to prevent them from taking place. A very useful one has already been devised and tested.

### The SR Grid

Back in the 1920's, a Polish mathematician, engineer, and student of human behavior, Count Alfred Korzybski, worked out a theory for control at the level of the elements of our experience. He spent the rest of his life trying to help people recognize the value of his discovery. Later, Dr. J. S. Bois developed an observing grid based on Korzybski's theory. He tested it and found that it yielded results beyond his expectations. He used Korzybski's terminology, since we always need a new word for new ideas. (Think of talking about electricity without using special words like volts, amperes, solenoids, capacitors, etc.) He called the observing grid the "Semantic Reaction Diagram." I shall describe this grid, and from now on we shall refer to it as the "SR Grid."

The two part body-mind grid gives us a crude breakdown of behavior in terms of "parts" of ourselves or others to observe. The "SR Grid" focuses on the *activities* going on within a person and around him. In this way it is like looking at the operational functions of a company, such as production, sales and finance, rather than looking at the material facilities of a company, such as the plant, the warehouse, and the office. The "SR Grid" is made up of seven sections. It allows us to become aware of seven general areas of closely interwoven activities going on within and around us at any given moment.

These activities are called: Thinking, Feeling, Moving, Electro-Chemical, Past Experience and Expectations. In this chapter I shall describe five of these areas and leave the other two to later chapters.

### Thinking Activities

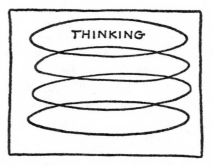

The Thinking area covers all our activities involving words and symbols. We use symbols when we talk, listen, read, write, plan, analyze, think, judge, design, and so on. Every time we go into a new area of knowledge we must learn a new set of symbols. If we wish to follow the market, we must learn the ticker-tape symbols. To play music, we must learn musical symbols. To do engineering design, we have to learn mathematical, mechanical, electrical, and perhaps chemical symbols.

As man has opened up new fields of study he has created hosts of symbol systems: meterological symbols, astronomical, biological, archaeological, not to mention such long-standing symbol systems as monetary, trade, religious, national, club, and military symbols. These symbols are signs to which we attach an arbitrary meaning. We think with them. We share our experiences—that is, we communicate with them. We store our knowledge, and accumulate it with symbols. Find out what symbols a man can use and you'll know his interests, what he has done, and what he is likely to be able to do. In Thinking activities, we would principally watch what he says and, if available, what he writes.

*Feeling*

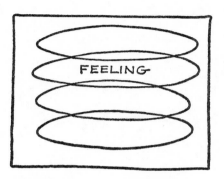

The area of Feeling represents the diffuse range of emotional experiences to which we refer when we use such words as enthusiasm, joy, anxiety, uncertainty, guilt, hostility, defensiveness, embarrassment, humiliation, love, fear. Our feelings are associated with our control and release of energy. They could be thought of as the carburetor of our system. They direct us towards or away from things; move us quickly or slowly. We can observe our own feelings directly. Sometimes we can sense another person's feelings, if we happen to be sensitive to them. Most of the time, however, we make guesses about how the other person is feeling on the basis of what we hear him say (his Thinking activities), and what we see him do (his Moving activities), about which I will talk in the next paragraph. Our Feeling activities are closely interlocked with our Thinking activities. No matter how objective, how rational we think we are at any time, feelings are present. Even when we're wild with rage or madly in love we keep on thinking to a degree. Both activities are present in every moment of our existence, and continually interact and change one another. Notice how the ellipses in the SR Grid overlap—a reminder to us of this interdependence of all of our activities.

*Moving Activities*

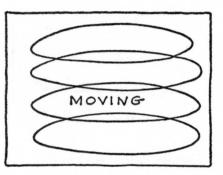

Our Moving activities include all our body movements and muscular patterns—the way we sit, hold our head, eat, walk, move our arms and legs. Other Moving activities would be facial expressions, body mannerisms, our breathing, the workings of our heart, lungs, stomach, liver, and kidneys. By changing what we do we can change how we feel and how we think. Our Moving activities, too, are a part of every moment of our existence.

*Electro-Chemical Activities*

We can usually observe our Electro-Chemical activities only indirectly through instruments. It is the level of body chemistry, of metabolic changes, of electrical impulses in our nerve

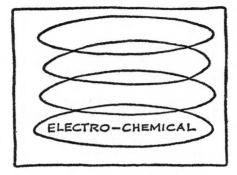

fibres, of physical changes at the molecular level. It can be observed to a degree with such instruments as the electrocardiograph, the electroencephalograph, which records our so-called "brain waves," the electromyograph, which records changes in muscle tensions, and the lie detector. The Electro-Chemical activities are affected directly by food, beverages, alcohol, anaesthetics, bacteria, hormones, vitamins, radio-activity, tranquillizers, and the like. A change by one of these agents can bring about profound changes in our capacity to think, feel, and move.

### Environment

The Environment includes all the goings-on in our immediate surroundings: the temperature, humidity, air pressure, noise, odors, natural surroundings, or room furnishings, light, color, and so on. Our Environment also includes all of the people around us and what they believe in—our bosses, fellow employees, neighbors, friends, families, relatives, customers, etc. Other factors in Environment are our church, school, clubs, customs, methods and procedures, company policies, laws, organization structure, economic and international condi-

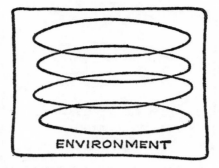

tions. All these factors interact on and change in varying degrees our Thinking, Feeling, Moving, and Electro-Chemical activities.

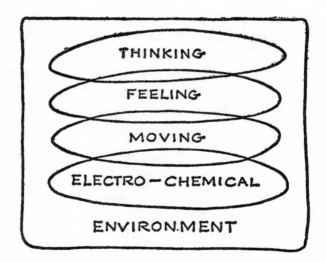

In the following chapter you will see how to use the SR Grid to locate many sources of negative factors within and around you. You will also see how you can eliminate many of these sources before they can drain away too much of your energy.

**Chapter 9**

# Shield your
# energy

*As you increase the number of moments throughout the day that you check in on yourself, you create an ever-more-effective shield against energy losses.*

## Managing by the Minute

When we are preoccupied large chunks of our time pass by unnoticed. We allow stress to build up steadily all morning and then "take a break" by relaxing at lunch. We drive hard throughout the afternoon meeting to get our points across and seek relief in a drink or a rest before dinner. Our fatigue at the end of the day shows that more energy has leaked out of our system than we realized as we went on with our work. By the time we wake up to this fact it's too late to do anything about it.

To break through the time barrier, we need to deal with

momentary impressions of our own activities and the happenings around us. The SR Grid gives us a systematic way of doing this. We can check in on each area of the SR Grid much as a watchman checks in at each of his call stations. We look for indicators of negative factors in our Thinking, Moving, and Electro-Chemical activities as well as the conditions in our Environment that create negative factors. As we increase the number of moments throughout the day that we check in on ourselves, we create an ever-more-effective shield against energy losses. This self-observation I shall call "managing by the minute."

Some people ask: "If you manage by the minute, you'd be constantly thinking of yourself. Isn't that what we mean by the word selfish?" Not at all. In fact, it's quite the opposite. When you manage by the minute you make frequent checks on yourself and your surroundings. Because you can catch many negative factors before they do much damage, your energy is high, your decision-making capacity is at its best, and your ability to grasp quickly the essentials of the situation is at its peak. Accordingly, you are in the best possible position to create on your own behalf and to help others. On the other hand, when you check in on yourself only every few hours, you are usually out of balance. The more you are out of balance, the more time you give to thinking about yourself and the more you tend to disturb others.

We all know that a seriously disturbed person is blind to his own negative factors. He blames everything and everybody for his troubles. It takes a good deal of patience and skill to get him to take even the quickest glance at himself. In the same way, when we're upset we tend to be unaware of our defensiveness, criticism, anxiety, and fears. The negative factors amongst our Feeling or emotional activities are usually so well concealed that we can discover them only by indirect methods. What, then, are some of the indicators of negative factors in our Thinking, Moving, and Electro-Chemical activities as well as in our Environment? Since negative factors, low energy and preoccupation go together, these will be indications of preoccupation in ourselves and others as well.

### Indicators in Our Thinking Activities
### of Negative Factors and Preoccupations

What are some of the ways we think and talk that go along with feelings of defensiveness, criticism, fear, anxiety, and other negative factors? Among the clues we can look for in our Thinking activities are the words spotted in on the following SR Grid:

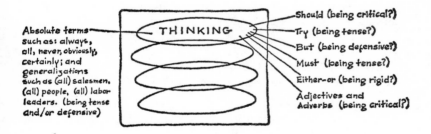

No one of these words in itself is "bad" or proof of energy loss. All of them have their proper uses. When I spot myself using one, however, I look to see if it is in a cluster with others. When this is so, I find that negative factors are present.

Another group of indicators I'm ever on the watch for contains what I call "inner concerns." By "inner concerns" I mean worrying about what other people might be thinking of me.

They crop up in questions such as "What impression am I making?" "What will the neighbors think?" "Is this the *proper* thing to do?" Associated with these inner concerns are various forms of faking—pretending I have skills, friendships, and possessions that are non-existent. Whenever this happens it takes all of my attention just to keep up the front. All of this is quite pointless as I seldom fool the people who are supposed to be impressed.

More subtle still are the parts I catch myself playing when I'm not aware. Like Walter Mitty, I sometimes secretly revel in imagining myself in different roles of power. By watching carefully over several years, I have been able to recognize several of these personalities that recur with regularity. By giving them names I can unmask them (*persona* is the Latin word for mask) and render them impotent with the ridicule implied in their titles—the Tycoon, the Mastermind, and Moses. All of these forms of deceit leave me vulnerable to criticism, keep me on the defensive, and rob me of energy.

### Indicators in Moving Activities
#### of Negative Factors and Preoccupation

As I have come to be able to observe my body, I have been surprised at how much it can tell me. My facial set, the position of my head relative to my body, the way I hold my hands, the way I'm sitting, what I'm doing with my feet, the way I'm breathing, and so on. I found that I didn't need a book to interpret these gestural patterns—they were only too obvious. At the top of page 67 are a few clues in my Moving activities I look for in order to discover negative factors.

Of these indicators I pay particular attention to various forms of talking and to tension. When I first began increasing

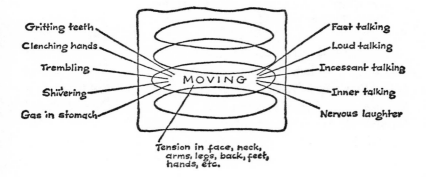

my awareness of my Moving activities, I found that I was talking to myself almost continuously. As I cut down on this inner gabbing, there was considerable drop off in muscular tensions. Speaking slowly and deliberately also served me well as a means of keeping alert and as a technique of relaxation.

### Indicators in Electro-Chemical Activities of Negative Factors and Preoccupation

When it comes to observing my body processes I'm reduced to making a few guesses. My doctor can give me a good deal of help in this area through his use of blood tests, urinalysis, basal metabolism tests, blood sugar tests, and so on. Apart from professional advice, I watch my diet, and don't dump unreasonable quantities of hard-to-digest foods and drinks down my gullet. Electro-Chemical activities are on the frontier of current medical research. I imagine that the next ten years will see a whole range of possibilities presented to us for extending our life span and increasing our energy resources through closer control of our diet and Electro-Chemical balances.

### *Elements in Our Environment that Breed Negative Factors and Preoccupations*

In general, I attempt to cut down the number of distractions in my environment. What is distracting to me, of course, may not bother another. Furthermore, distractions vary according to one's location, feelings, and condition. Some potential sources of negative factors I watch for as I study my environment are:

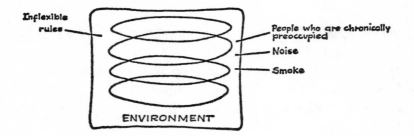

The greatest source of negative factors in our environment are people who are chronically preoccupied: people who talk incessantly; people who argue, who criticize, who complain. Whenever I am with such people, I place most of my attention on staying alert. I ask thought-provoking questions and avoid stereotyped conversation in an attempt to keep them alert too.

### *Shielding Your Energy*

Let us assume now that you've checked in on the various areas of the SR Grid and have discovered some indicators of negative factors. What can you do about it? There are a number of possibilities, depending on the sort of clues you've found. If you catch yourself using some of the clue words the answer is simple. Stop using them. If you discover yourself thinking something critical, something defensive, playing a phony role, or the like, turn your attention away from these things immediately. Force yourself to think of something else. If the thoughts keep recurring despite your best efforts, you can then try changing some of your Moving activities: go for a walk, do some exercise, practice some skill, concentrate on observing something in detail. You can also have a bite to eat, drink some coffee or other beverage, or even take a nap. Sometimes you can help yourself by changing your environment temporarily, or, if the negative factors are very persistent, make a change in your place of residence, your job, or your country. *Remember that a change in any activity will change all other activities.*

You will find it useful, from time to time, to sit down with a diagram of the SR Grid in front of you and to write down a list of some of the possibilities for change that are open to you in each area of the Grid. You will be surprised at the tremendous range of possibilities for action lying within your area of freedom, even in the poorest circumstances.

The simplest way to both shield and develop energy is to sleep. When you sleep you are free from desires. Your body is left undisturbed to restore the balances you've upset during the day. After a long night's sleep, too, you will find it much easier to cope with negative factors.

Finally, remember that your goal is to win back alertness and gain time. Like the Romans, who had to fight continuously to extend and hold their frontiers against the barbarians, you will have to forever struggle against preoccupation and negative factors. When you lose some ground, as you will from time to time, avoid criticizing yourself. Capitalize on your withdrawal instead, by letting it build your resolve not to give ground in that area so easily the next time.

**Chapter**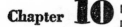

# The birth
# of vision

*The continuous invention of new ways of observing is man's special secret of living.*

*—J. Z. Young*

### The Blind from Birth

The third phase of increasing the clarity, accuracy, and speed of your picture making involves broadening your knowledge and experience. To understand why this is so you would need to seek the answer to this question, "What controls the kind of impressions I get of what is going on about me?" If we could question a newborn baby about what it sees, hears, tastes, and feels, we might be close to the answer. Unfortunately, we have to wait a good many years before the child can give us clear answers, and by that time it's too late for our purposes.

A way has been found, however, to study what happens when a person first uses his eyes. There are certain kinds of congenital eye defects that are blinding but can be corrected with surgery. In this century a number of people, blinded from birth in this way, have been successfully operated upon. They have told researchers in detail about their first impressions of the world of sight.

Take the case of Tom Morrisson. Tom is eighteen years old and has been blind since birth. He graduated from high school with honors. When we meet him he is lying propped up in a hospital bed in a private room of the Mayo Clinic. A week before he had undergone surgery, and this morning the doctor is coming to remove the bandages from his eyes for the first time. The doctor comes into the room, questions Tom, and gently unwinds the surgical gauze. He peers into Tom's eyes and proclaims them optically perfect. Tom is facing a white, painted chest of four drawers standing against the wall at the foot of his bed. To his right is a window through which the mid-morning sunlight is diffused by the tilted venetian blinds. We step over to the bedside and ask: "Tom, what do you see?" What do you think his answer will be?

I have put this question to managers in many discussion sessions. The speculation ranges far and wide. Some say Tom would only see lights; others say he would see the chest of drawers. Many say that, though it would take a little time, gradually Tom would make out the various items in the room. "After all," they say, "you told us that he had touched everything in the room when he first came into it. He would therefore already know what was in it."

### Out of the Dark

When interest and curiosity have reached a peak, I take a

copy of Professor J. Z. Young's book, *Doubt and Certainty in Science*. In this book, Professor Young, an authority on the human brain, describes what a man blind from birth would see on first opening his eyes:

> The patient on opening his eyes for the first time gets little or no enjoyment; indeed, he finds the experience painful. He reports only a spinning mass of lights and colors. He proves to be quite unable to pick out objects by sight, to recognize what they are, or to name them. He has no conception of a space with objects in it, although he knows all about objects and their names by touch. "Of course," you will say, "he must take a little time to learn to recognize them by sight." Not a *little* time, but a very, very long time, in fact, years. His *brain has not been trained in the rules of seeing.* We are not conscious that there are any such rules; we think we see, as we say, "naturally." But we have in fact learned a whole set of rules during childhood.
>
> If our blind man is to make use of his eyes he, too, must train his brain. How can this be done? *Unless he is quite clever and very persistent he may never learn to make use of his eyes at all.* At first he only distinguishes a mass of color, but gradually he learns to distinguish shapes. When shown a patch of one color placed on another he will quickly see that there is a difference between the patch and its surroundings. What he will not do is to recognize that he has seen that particular shape before, nor will he be able to give it its proper name. For example, one man when shown an orange a week after beginning to see said it was gold. When asked, "What shape is it?" he said, "Let me touch it and I will tell you!" After doing so, he said that it was an orange. Then he looked long at it and said, "Yes, I can see that it is round." Shown next a blue square, he said it was blue and round. A triangle he also described as round. When the angles were pointed out to him he said, "Ah! Yes, I understand now, one can *see* how they feel." For weeks and months

after beginning to see, the person can only with great difficulty distinguish between the simplest shapes, such as a triangle and a square. If you ask him how he does it, he may say, "Of course if I look carefully I see that there are three sharp turns at the edge of one patch of light and four on the other." But he may add peevishly, "What on earth do you mean by saying that it would be useful to know this? The difference is only very slight and it takes me a long time to work it out. I can do it much better with my fingers." And if you show him the two next day he will be unable to say which is a triangle and which a square.

The patient often finds that the new sense brings only a feeling of uncertainty and he may refuse to make any attempt to use it unless forced to do so. He does not spontaneously attend to the details of shapes. He has not learned the rules, does not know which features are significant and useful for naming objects and conducting life. Remember that for him previously shapes have been named only after feeling the disposition of their edges by touch. However, if you can convince him that it is worth while, then, after weeks of practice, he will name the simple objects by sight. At first they must be seen always in the same color and at the same angle. One man having learned to name an egg, a potato, and a cube of sugar when he saw them, could not do it when they were put in yellow light. The lump of sugar was named when on the table but not when hung up in the air with a thread. However, such people can gradually learn; if sufficiently encouraged they may after some years develop a full visual life and be able even to read.

### Principles of Observing

You, like Tom, have not only to learn to see, but to hear, touch, taste, and smell. In your earliest infancy everything was one big blur. Once you caught your breath after the first slap on your behind, you began to experiment. Every waking moment was filled with looking, tasting, smelling, touching, and listening. It was a long process with plenty of smiles and tears along the way. By the time you learned your first words you had already collected an enormous file of models in your brain. With these models, these bits of sensory data, you could form impressions of some of the happenings about you. In areas where you had many models you could form mental pictures that were clear and representative. They allowed you to act effectively and with confidence. In all other areas you were lost. You were forced to grope around until you learned the necessary models to allow you to "make sense," to make clear pictures of the situation.

Here then are some of the basic principles of observation:

1. You are the creator of your own unique elements of experience, therefore you can change them at will.

2. You select and arrange the parts of your sensory impressions on the basis of your past experience.

3. As your experience in any area grows, you can see more in that area (make clearer, bigger, and more accurate mental pictures of what's going on).

4. Since your experience is never complete, you never can perceive *all* of what's going on; you always leave out a great deal in your observations.

These principles can be summarized in a diagram, like this: *

---

* This diagram is an adaptation of Alfred Korzybski's "Structural Differential." See Korzybski, Alfred, *Science and Sanity* (Lakeville, Conn.: The Institute of General Semantics).

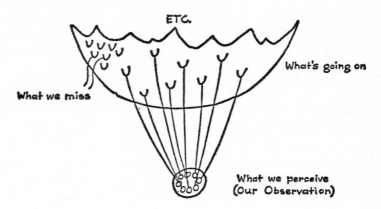

What you perceive or abstract from what's going on you make into a mental picture. This abstracting process is shown in the diagram as a series of lines running down from the parabola of "What's going on" to the circle of "What we perceive." The "ETC" at the top and the lines running from the left side of the parabola are to remind you that in making your selection from "What's going on" you are always forced to leave out most of what is happening. Remember the man blind from birth. Like him, you can only see what you have trained your brain to see. You, like everyone else, in one lifetime can only learn a bit of what there is to know, hence, you can only see a bit of what is going on.

The diagram can be extended as shown above to indicate what you usually do after observing something.

You describe your experience either to yourself (thinking), or to someone else (talking or writing). When you try to describe what you experienced, you are forced, no matter what you do, to leave out the bulk of what you observed. If you are doubtful about this you can check it for yourself in this way. Look out a window for two seconds. Now take some

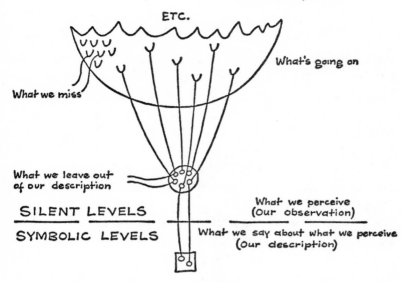

paper and write down a description of every detail you can
recall seeing. Even if you have a poor recall it will probably
take you four or five minutes to do this. Note! 4 or 5 minutes,
or 240 to 300 seconds to describe a two-second look! There
just isn't enough time to fully describe most of your observations to the people around you. Accordingly, most of the time
you select a fraction of one per cent of what you observed and
give this to your listeners as "the facts."

Look at the chain of information losses that accumulates in
even a short conversation:

Loss 1: You miss most of what is going on when you
make your initial observation.

Loss 2: You leave out most of what you did see when you
give your description to your listener; often you
unknowingly leave out some critical elements of
the happening.

Loss 3: Your listener only hears a part of what you say.

Loss 4: Your listener can only describe part of what he heard.

Now dwell on the plight of the manager. He has to make his decisions for the most part on the basis of information given to him by others. Since this information is about past events he can never check its accuracy directly. He can, of course, check other people's impressions of what happened, but, as you know, he can at best gain only an approximate picture. The task of a manager has been likened to that of a driver of an automobile who must steer solely on the basis of information given to him by a passenger. To make matters worse, the passenger can only give the driver the information he gets from the rear view mirror.

The difference between the pictures that various employees make of something as tangible as the company product can be striking. Here is an example, taken from Executive Action,* that vividly illustrates this point:

> To the president of the company, who had come in at a high level from another company, the product acquired its meaning from the fact that it was respected by certain reputable banking establishments and by his business colleagues and that it formed the core of an active enterprise with a sound future. To the sales vice president, the product was a gleam in the weary buyer's eye, a kindler of excitement at sales conventions, an impressive thing which could be set up on display tables or even shown with pride on the stage of an auditorium. To him it was a passport to acceptance in the marketplace. To the controller, the product was one of the chief variables in the great game of figures he played, and he did his best to keep it where he wanted it so that the figures would come out right. To the

* Learned, Ulrich and Booz, *Executive Action* (Cambridge: Harvard University, 1951).

manufacturing vice president, it was the embodiment of a long and careful process which he must supervise. To the purchasing agent, it was the raw materials in their finished form; to the machine hands, it was a pliable but sometimes stubborn kind of physical substance whose shape they had power to change; to the maintenance man, it was a microscopic measure of wear on the bearings. In the most literal sense, each person *saw the product in his own way.*

Keeping these examples of information losses and individual differences in perception in mind, you can appreciate the value of these rules for conserving your time.

1. If it's important that your information be accurate, check personally and then get as many viewpoints of others as time will allow.

2. Even when you have checked the "facts" carefully, be prepared for the unexpected. Keep your plans sufficiently flexible to admit new data.

3. When someone fails to turn up for an appointment; when your customer seems downright unreasonable; when your subordinates make "stupid" mistakes, reserve your judgment. Conserve your energy and look for more information to explain the incident and ways of preventing its recurrence.

### Failure in Action

Once you understand that we have to learn to see, you can better appreciate the dilemma of a man who has been given an order he fails to understand. He may feel a great urgency to act but the questions he asks himself are: "How do I act?" and "How fast should I go?" As we were in the example of the last chapter, he is faced with a blurred picture on his screen, and he is filled with uncertainty. If pressed to act, he'll be overcome with fear. He'll sweat, tremble, but won't act. Forcing him to do so is like shouting "Run!" at a blind man in the middle of a busy intersection.

Picture a department manager who has graduated by seniority to his present position during a 10-year boom period. Suddenly there is a slump in sales, and his division manager calls him for a heart-to-heart talk. Here's how the conversation might run: "Joe, you've been doing a mighty fine job in your department. Had no complaints at all. Now, as you know, however, our inventories are piling up and the G.M. has been after me to cut costs. I know you can streamline your operation easily. Get together with your boys and start going. If you need any help, just let me know. One thing, remember, I'm expected to produce results, and fast!"

Joe has always managed to get out the production. Since his company has been operating in a seller's market for a long time, the focus has always been on production. Along the way, Joe has never been under any boss who took time to show him how to view his job as a whole. Always it was just a matter of keeping up the flow of goods, looking after his men, and maintaining quality controls. Now he's faced with an ultimatum. Sure, the boss said he'd be willing to help, but what help was Joe going to ask for? He couldn't *see* what more to do, and, of course, didn't know what questions to ask. Imagine, now, one month later, after an almost uninterrupted series of sleepless nights, how Joe feels when he's called in to explain to the boss why costs aren't down. I'll leave you with the picture, and let you fill in the outcome from your experience.

Take another case, not quite so dramatic but one that frustrates many sales managers  Bill Farrell is one of the company's best salesmen. He covers a key area and manages to beat his quota by a wide margin every year. Trouble is that Bill won't fill in his sales reports. On occasion, when pressed hard enough, he will send them in for a month or so, but most of them are obviously completed in a hotel room. The sales manager has a special group of market analysts who are trying to discover emerging trends in the market and make a five-year projection. He has high hopes for the outcome of this project. But Bill blocks them all. He simply can't see why those damn fools in head office don't keep their paper work to themselves. "I'm making the sales, ain't I?" he says. "Then why don't you leave me alone, and not try to make a paper jockey out of me?" The sales manager has explained the reasons to Bill at least twenty times but to no avail. Bill has no models, no experience in sales analysis to help him make the appropriate picture. Without a picture, he can hear the sales manager's words, feel the sales manager's sense of urgency—but he won't act. He won't act, at least, as the sales manager wants him to. He can't act in this way unless he has a picture similar to the one the sales manager has made for himself.

Other illustrations come to mind: The president with experience largely in sales who couldn't see why the manufacturing department wasn't able to switch over, in the middle of a big production run on one model, to another model in a week. The Director of Research who fails to see why the Executive Vice-President won't allocate him a budget and let him carry on in the way his researchers feel is best. The office manager who forecasts that the new order processing procedure won't work. He can only see the old, familiar way of handling orders and invoices. The list could go on, as you know, indefinitely.

Let me call special attention to one statement of the quotation from Professor Young: "The patient often finds that the

new sense brings only a feeling of uncertainty, and he may refuse to make any attempt to use it unless forced to do so." It's
little wonder that many people balk at a major method or tool
change; that a customer, when presented with a dramatic innovation in your product, may turn aside a sales presentation
even though the performance figures are spectacular; that a
man in an art gallery can get downright indignant about some
contemporary abstraction and refuse to listen to explanations
of its technique and purpose. Without appropriate models, the
necessary atoms of experience, these people are blind to the
possibilities of the new event. Furthermore, since they're at
a loss, they feel uncomfortable, they struggle to get back to
the things they know and can handle with surety.

*Training Your Brain*

You are living in a world that research is changing by the
hour. It is no longer enough for a man to learn a trade, get a
degree, or achieve a managerial position and rest on it. It's
small consolation to be a crack buggy driver in an age of auto-
mobiles. To act and to keep acting effectively, you must have
mental pictures that are clear and representative of the changing situations around you. Since the rate of change is accelerating too, you need to be able to make not only new pictures,
but more and better pictures. You can only do this by continually training your brain.

Your special secret of living, as Professor Young says, is
"the continuous invention of new ways of observing." In the
following chapter you will see how to build an investment
portfolio for growth. As you broaden the range of your knowledge and experience, you will be able to improve the accuracy
and speed of your picture making. This in turn will increase
your output and help you to overcome time pressures.

Chapter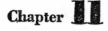

# Invest in growth

*Happiness is neither virtue nor pleasure*
*nor this thing nor that but simply growth.*
*We are happy when we are growing.*
—WILLIAM BUTLER YEATS

### *Your Opportunities for Investment*

You know that it is within your power to overcome time pressures. You have two working principles to guide you in meeting this challenge. One of these relates time pressures to your output. The other states that you can increase your output as you refine your techniques of self-management.

You have looked at a method that will allow you to refine the smallest manageable elements of your experience. You have seen that this method of improving the clarity, accuracy, and speed of your picture making has three parts; namely, increasing your alertness, increasing your available energy, and, finally, broadening your knowledge and experience. This method of overcoming time pressures could be summarized in this way: *keep developing your personal resources and you will have enough time to do what you want to do.* In this chapter you can review your opportunities for investment in growth. You will find, too, a suggested policy for investment that should give you a satisfying yield in the growth of your personal assets.

Your opportunities for investment can be grouped into five kinds of activities. These are: Exploring, Training, Conforming, Sleeping, and Idling. A well balanced portfolio for growth will include investments in all of these five activities. Only you can tell what combination best fits your needs. To help you in your selection, here are some details on each of the areas:

### Exploring

You can invest in exploration in two ways; by familiarizing yourself with new ideas, and by trying out new sensations. Like the man blind from birth who was learning to see, every new experience you have increases the range and speed of your picture making.

You explore the world of ideas when you read books, listen to people describe their ideas and experiences, attend plays, and formulate new methods and principles. You expand your knowledge when you learn to work with or deepen your understanding of symbol systems, whether they be linguistic, mathematical, musical, pictorial, engineering, or diagrammatic.

As your business grows in complexity it touches on more aspects of community and world affairs. There was a time, not so long ago, when a business man considered it good practice to give his full attention to the mechanics of his business. Now a manager can no longer afford to limit his knowledge to business alone. New trends in marketing call for some understanding of psychology, sociology, and related sciences. The pressures of foreign competition make it useful to have some grasp of cultural anthropology and political science. The rate of technological progress calls for ever longer range planning. This in turn requires a growing understanding of statistical and mathematical techniques. It calls, too, for an understanding of historical processes, so that the lessons of the past may be built into the predictions and plans for tomorrow. If you lack knowledge in these and other areas you will be blind to key factors on which the future of your business will depend.

You explore the world of sensations when you try out new sights, sounds, tastes, or feelings. You broaden your "know-how" as contrasted to your "know-about" when you go to new places, eat new foods, speak new languages, listen to unfamiliar music, look at new art forms, and meet new people. To use the analogy of a radar-electronic computor installation, when you explore new sensations you increase the sensitivity and range of your antennae. When you explore the world of ideas you increase the capacity of your computor. Your adventures in the worlds of sensory experience and ideas therefore complement one another. Too much time invested in sensation will put you out of touch with new trends and developments in business and world affairs. Too much time spent in talking, planning, and reading will surely dub you as an impractical theorist.

The more novel the experience or ideas you explore, the greater will be the risk you take. On the other hand, the extra risk and effort of exploring far afield will often yield you rich treasures in growth. As you continue to invest in exploration, you will train your brain with greater ease. You will find, also, that as your capacity grows you will see an ever-widening array of opportunities for creation, satisfaction, and growth.

### Training

As you have seen, the more you explore, the larger will be your empire of knowledge and experience. With the expansion of your holdings you will need increasing skills to take advantage of all your newfound opportunities. If you fail to increase your capacity to concentrate, observe, communicate, and relax you will soon be worn to a frazzle trying to keep control of your range of widespread interests and activities.

In Chapter 6 you saw what you can do to train yourself to become more alert. The daily practice period will not only help you overcome preoccupation, but will give you a growing power "to do." As you know, if you are to keep control of an expanding range of interests, you must match these with a growing capacity to discipline yourself.

In the next part of this book you will find seven different skills described that will help you increase your output and manage your time. Rather than attempt them all at once, take a tip from Benjamin Franklin. In his autobiography, Franklin describes how he acquired various skills by concentrating on the practice of one skill each week.

Training calls for a heavy investment of energy at first. It demands that you discipline yourself to practice regularly whether you feel like it or not. You know, however, how much satisfaction can come from the increase in your skills that training can produce. This satisfaction seems to increase the more heavily you invest in training.

## *Conforming*

To be accepted by your family, your company, your fellow club members, and your neighbors you have to conform to their "ground rules." Most of the investment you make in learning to conform was made during your childhood. Your parents, teachers, and playmates forced you to conform. At that time your investment was heavy but it tapered off as you became "civilized" and accepted the role of an adult.

When you grew up you came to realize the strange character of conforming. Whereas during childhood it took a lot of effort to conform, in adulthood, quite the reverse, it takes less effort to conform than to do something original. If you fall prey to inertia you tend to avoid the effort and responsibility of innovation. You come to operate largely on habit, and escape from the unpleasantness of community problems, the complaints of subordinates, the anxieties of delegation, and the irritations of product and method improvement.

Like everyone else who wants companionship in life, you must conform to the extent of social amenities. If you make it a habit to conform, however, you quickly lose your freedom to manage your time. Imperceptibly at first, and then ever more swiftly, you stagnate. One day you wake up to find yourself being discussed as a "problem."

## *Sleeping*

When you sleep your system can devote its entire attention to restoring the balances you have upset during your day's activities. Sleep is the only method you can use without any training to free yourself swiftly from desire. When you are making large investments of your energy in other activities, you would be well advised to invest heavily too in sleep. For a very small investment in sleep you will gain a large return of energy and a sense of well being.

## Idling

One of the rewards of a well-balanced investment portfolio is the opportunity to give over a little of your resources to idling. Probably from no other investment can you get such sheer pleasure. We all know the joy of letting go after a hard physical workout, just lying back and not giving a damn about anything. You enjoy idling also when you sit on the patio and lazily daydream. You want nothing. You think about nothing. You do nothing.

You have the opportunity to invest in idling and its joys to the degree you are managing your time effectively at other times.

## Obstacles to Investment in Exploration

At this point you may be looking forward eagerly to investing in ways to increase your growth. To conserve your resources, however, I would caution you to be on the lookout for certain resistances that can develop within you quite unawares. Having acknowledged them you can take appropriate steps before they grow unmanageable.

To understand the nature of these obstacles, consider a typical anthropological situation. You can do this by joining in your imagination a party of cultural anthropologists who are being paddled up a river in the rain forests of New Guinea. They have chanced on a native who has volunteered to lead them to a hitherto unknown tribe living near the headwaters of the river. Let's assume that they have made contact with the villagers and have achieved sufficient acceptance to be allowed to live with them for a while. They want to learn as much as possible about the customs of these people: how they hunt, eat, dance, marry, worship, and so on. You have been assigned to study the habits and customs of one young native hunter of the village. His name is Nanga.

You live with them for a month and study what Nanga, his wife, and his two children do. You learn a good deal. Among other things, you learn that Nanga always sleeps on his left side, facing the door of his hut. All the hunters of the village do so. It is taboo, or bad, to sleep otherwise. Nanga's wife always stirs the food in her cooking pots in a clockwise direction. No other way will do—it will put a taboo on the food should she stir it any other way. When Nanga goes hunting he always carries his spear point up—it is bad or taboo to do differently. If he should meet a stranger in the jungle, he must always attack. Anyone strange is bad, an enemy. In fact, you notice a common pattern running through Nanga's behavior. He has a range of things he knows and does, always in the same way. These are good. Anything he does not know directly or through knowledge of the village group is automatically bad and must be destroyed or avoided at all costs.

For Nanga and his fellow villagers this is survival behavior. It works very well, too, since primitive peoples live in an environment that, for all intents and purposes, can be counted on to be the same, century after century. All Nanga has to do is to learn one set of rules during childhood and he's fixed for life.

Keeping in mind the story of Nanga, ask yourself this question, "What is likely to be my first, unthinking reaction to something strange?" Before you answer, remember the reaction of the man whose sight was restored. On first opening his eyes he found the experience painful. Now, what might be your first reaction when offered a platter of fried grasshoppers or a piece of fresh raw blubber? If you're like most people, your first reaction will be to make a judgment—a negative one—ugh!, bad!, or the like. As soon as you do this you trigger off a pattern of retreat. You back away from the new experience.

### Invest in Growth

Every day you see many things going on that demand no action on your part. Some of these things are strange, and your first reaction is to judge them negatively. This is most likely to happen when you are not alert, when you are preoccupied. But look what you have done! Your native resistance has blinded you to what was new in the experience and robbed you of an opportunity to grow by exploring it.

Most opportunities for growth lie in the unfamiliar. Study with interest the people, products, and methods that seem strange to you. If you have the opportunity to visit foreign lands, search for the unfamiliar rather than for things "like they are back home." When you see a product of foreign make or a domestic one that seems unique, examine it carefully and try to understand how it has been made and what its characteristics are. Keep relating these observations to your own work. Keep asking yourself this question, "How could the 'know-how' built into this product or method be applied in my work or recreation?"

Some of your experiments and investigations into the unfamiliar may prove unsuccessful or even unpleasant. When you invest your time in this way, however, you can expect a good return on most of your ventures. Those that fail will be more than offset by the successful ones. Slowly at first, then ever more quickly your wealth of experience and knowledge will grow. If, on the other hand, you reject the risks, you are left to scrabble for the leavings alongside the docksides of the familiar. If you follow an investment policy that backs exploration—exploration of the world of experiences and the world of ideas—you will grow, and, as William Butler Yeats puts it: "We are happy when we are growing."

# SUMMARY OF PART II

1. You can increase your output as you increase your capacity to get accurate, clear and fast impressions of what is going on around you.
2. To improve your mental picture-making capacity you need to increase your alertness, increase your available energy, and increase the range of your knowledge and experience.
3. To increase your alertness you must overcome your tendency towards being preoccupied.
4. You can gain more moments of alertness each day through changing your routines, practicing some skill daily and through cultivating interests centering on observation.
5. As you eliminate criticism, defensiveness and other negative factors, you will build up your available energy.
6. To save energy you must know where you're losing it.
7. You can create an ever more effective shield against energy losses as you increase the number of moments throughout the day you check on your activities at all levels.
8. Remember that at best you can get only a fraction of the information about what is going on. Check to find out what other people see in order to improve the quality of your impressions.
9. To increase the range and speed of your picture making you have to train your brain through exploring new ideas and sensory experiences.
10. To keep control of your expanding knowledge and experience, give some attention every day to increasing your skills.
11. Use the creative power of sleep. The more demands you make on yourself, the more sleep you will need.
12. Most of your opportunities for growth lie in the unfamiliar. Avoid judging the strange and give yourself an opportunity to grow through observation.

PART **III**

# Skills for
# Managing Time

Chapter **12**

# Knowing when to stop

*When you can't find an answer, stop, and save time by restating the problem.*

### *Opportunities for Increasing Your Skills*

As you gain more moments of alertness you will be free to invest more heavily in developing your skills. You will find that most of the opportunities for increasing your skills lie in your daily work. When you speak with alertness, you can both hear yourself and watch your listeners. You can refine your techniques of speaking as you watch their reactions as revealed by the changes in their facial and body sets. Similarly, as will be discussed in the following chapters, you can increase your skills while on the job in listening, writing, reading, diagramming, and in creating analogies. To the degree you make each working hour contribute to your skills, your capacity to manage your time will grow.

There is one skill in particular which will allow you to make large savings of time. I call this the skill of "Knowing When to Stop." As an introduction to what I mean let me use an analogy. In my part of the country we have snow for four months of the year. The motorist stuck in a snowdrift is part of the normal scenery in these months. Some people, when they first get in this fix, set their teeth and press down hard on the accelerator. Most times they only succeed in getting embedded even deeper in the snow. The faster they spin their wheels, the more gas, rubber, and good temper they burn up. These people have no skill in "knowing when to stop."

In a similar fashion, when you've worked on a problem for some time, yet seem no closer to an answer, it's a great time-saver to "know when to stop." If you keep grimly searching for a solution you can waste hours or days of time, exhaust your energy and lose your alertness, and, therefore, your freedom to manage your time as well. Of course, it's not enough just to stop. What I call the skill of "knowing when to stop" also includes *knowing what to do next* to break through your impasse. A key to this second step can be found in the principle of projection.

### The Principle of Projection

You, like the man blind from birth who was learning to see for the first time, have to train your brain through repetition. A vital part of this process seems to be the coding or classifying of the experience. As soon as you have a name or symbol to attach to your observation, it helps you to make and retain a sharper picture. A sharp picture gives you the choice of acting decisively. By translating your experiences into words you maintain a feeling of confidence.

Words sharpen mental pictures, but they impose limits on the user at the same time. Just as any area of land may be mapped in many ways (geographical, topographical, geological), so you can talk about your experiences in many ways. You can, for example, classify a man by his education, his work experience, his personality, his social habits, or his family life. Each classification allows you to point up certain features while neglecting others. Each word in a particular context is a mapping device. When you use it you direct your perception in a selective manner. In addition, you cannot help projecting your past experiences, related to the word, into the present you are talking about. This projection can be diagrammed in the following way:

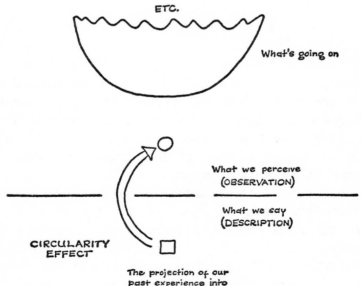

The projection of our past experience into our observation.

*Breaking Through to New Ideas*

The nine-point square problem demonstrates how words, through circularity effect, can limit us. Take a pencil and a blank sheet of paper and mark down nine dots in a square pattern like this:

      •      •      •

      •      •      •

      •      •      •

Now place the pencil on one of the dots of the square and draw four straight lines through all of the remaining dots without lifting the pencil from the paper or retracing any of the lines you have drawn. If you are not already familiar with the problem you will get more out of this chapter if you make an earnest effort to find the answer for yourself.

Did you find that you couldn't pass the lines through all the points? Most people who try this problem find that no matter what they do they always end up this way. When they see the solution (see below), they usually exclaim, "But the lines go outside the square!" Then they stop, think back for a moment to the statement of the problem, and realize that nothing had been said about not going outside the square. In fact, the word "square" was quite irrelevant to the problem. When they heard it, however, it triggered off a set of past experiences related to "squareness," and they projected these into the problem. So long as the word "square" dominated their thinking they were limited in their search and could never find a correct solution.

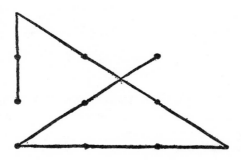

Whenever you are faced with a problem that seems unsolvable, whenever you wish to break through to new ideas, you can remind ourselves of the nine-point square problem.

This in turn will remind you of the next step in the skill of "knowing when to stop." This is the step of searching for your assumptions. Somewhere in your approach to the problem you are assuming something that isn't so. This idea you are taking for granted usually escapes your notice because it seems to you to be so obvious. Here is one way of discovering this self-imposed limitation on your thinking:

1. Write out a statement of the problem in simple language.

2. Using this statement as a guide, write down a list of as many of the things you are taking for granted as you can discover. Write down every assumption no matter how "obvious" it may appear to be.

3. Now go down your list and ask yourself these questions after reading each assumption:
   a) What if this is not so?
   b) What other alternatives can I think of?

4. On the basis of your questioning, write out alternative statements of your problem.

Here, for example, is the kind of design problem that students work on in the creative design course at M.I.T.:

"To design a multipurpose sanding machine capable of doing both rough and fine sanding."

Applying the method of searching for assumptions you might, if faced with this problem, write down a list like this:

I am taking for granted that:

1. There is no other person readily available who could do this design job more effectively than I can.
2. There is not a machine presently available that will do the job I want.
3. It would be advantageous to have a multipurpose machine.
4. The machine must perform with abrasives.
5. The final surface must be obtained by removing material from the rough surface.

After questioning these assumptions you might end up with a re-statement of your problem in this form:

"To design a multipurpose smoothing machine capable of giving a rough or smooth finish. Smoothing may be accomplished by either the addition or removal of material."

Many of the major breakthroughs in science, production,

sales, and finance have been the result of a re-statement of a problem or goal. For example, if you speak of "dust" as a fluid rather than as a solid, it opens up the possibility of pumping it through pipes instead of shoveling it into cars or trucks. If you speak of return on investment rather than profits, among other things it can allow you to discover unprofitable sales. When General Foods began thinking of their product "Minute Rice" as a vegetable in addition to regarding it as a dessert, they opened up a large new market for their product. When a bank begins to think of "money" as inventory rather than as an asset, it can then use inventory control techniques to increase its turnover of money.

### Knowing When to Stop

In the early stages of the development of a product or method you can usually get the greatest return for effort expended. Then, by the law of diminishing returns, you reach a point where only slight improvements can be achieved at high cost. Under today's condition of rapid innovation, you can waste a lot of time and lose your competitive position by carrying your improvements into the area of diminishing returns. It is small satisfaction to have the best darn ice box on the market.

Rather than improve the efficiency of a propeller, jump to a jet. Rather than buy another desk calculator for the accountant, look into punch card systems. More and more, the question "How can we eliminate it?" is superseding the question "How can we improve it?" The former question, you recognize, is part and parcel of the skill of "Knowing when to stop." With increasing alertness, you will have the opportunity of applying this skill more often. You will stop before you begin to improve something or when you can't find an answer, and save time by restating the problem.

# Chapter 13

# Your principal
# tools are words

*Strive for knowledge, delicacy, and precision in your use of words.*

### Working with Words

Your principal tools are words. You use them to give information, to gather information, to issue instructions, to plan, to analyze, and to control. Depending on your stage of business practice you use words in different ways. At the stage of the trader, a manager uses words in volume. Rather than blending the differing shades of word meanings, he tends to flood his listener with words. He keeps pouring them on until the listener either agrees or is able to interrupt with a counter-argument. At this stage the principal method of overcoming resistance is that of talking louder and faster in an attempt to wear down opposition by the weight and strength of the delivery. The use of words at the stage of trading is typified by the hawker mak-ing his spiel in the market place.

At the stage of calculated planning, a manager is conscious of more relationships and uses his words with some care. He often speaks, as we say, with diplomacy. He tries to avoid hurting the other fellow's feelings if possible. Talking and writing at this stage of business practice often assumes characteristic patterns that tend to appear at each level of management within a company. These modes of talking become almost a badge of the administrator. A manager can easily fall prey to these standardized statements, use them habitually, and lose his alertness in the process. A group of men talking jargon of this kind can slow a meeting to a glue-like consistency. Carried to an extreme, polished talking can degenerate into bureaucratic gobbledygook.

To manage at the stage of creative planning you know that you must be able to deal with your smallest elements of experience. Now you enter the shadowy area of possible meanings and reactions that can only be partially mapped by techniques like that of the SR Grid. When you utter a word you know that you set off a reverberating chain of thinking, feeling, moving, and electro-chemical reactions within yourself and your listeners. Like a surgeon probing among the billions of neurons in the brain, you need to work with delicacy and precision. You recognize that words are the master tools that control your creative capacity.

To guide you in your use of words as precision instruments, you can use three principles. I call these the principle of complexity, the principle of distortion, and the principle of projection which has already been discussed in the previous chapter.

*The Principle of Complexity*

On the surface, the process of communication looks like a simple matter of transmitting and receiving a message. The chief problem appears to be one of selecting the proper words. This view of communication could be diagrammed this way:

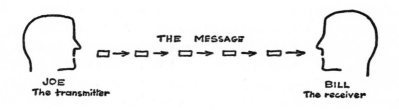

Treating communication as merely a matter of words can lead to time-consuming misunderstandings and errors. Here, for example, is a way of taking more factors into consideration: Think of a conversation between Bill and Joe as an interaction between their Thinking, Feeling, Moving, and Electro-Chemical activities in an Environment during a period of time. This trans-action could be diagrammed like this:

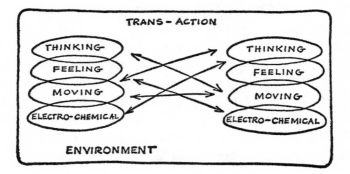

For example, Joe could be saying to Bill, "You're doing fine." This statement is the part of this thinking activity that Bill can observe. Privately, Joe might be thinking that Bill is frustratingly slow. If so, he is broadcasting his feelings of impatience and irritation, and Bill will pick up this message much clearer than the words. Furthermore, Joe will show revealing signs of tension in his body sets and in the coloring and inflection of his speech. Although Bill will probably not directly interpret these revealing clues in Joe's Moving activities they will contribute to his over-all impression of Joe's insincerity. The time and place (Environment) that Joe selects to make his statement also contributes meaning to the transaction. If he said it as he was hurrying out the door on his way to a golfing date, it would have a different impact than said quietly in the privacy of his office. When you are alert you can pick the most appropriate time and place to make your remarks. If you do this most of the time you will be known to have "good timing."

### The Principle of Distortion

You saw in Chapter 10 that you are limited in your observing. Like the man blind from birth you are limited by your range of past experience, by your personal makeup, by your position, by your purposes, by your background, and by your energy. One consequence of this is that whatever you say, write, hear, or read is always distorted to a degree. In the first instance, when you observe you leave out almost all of what's going on. Then, when you come to report on your observation you leave out the larger part of that too. In practice, the principle of distortion can be stated as a series of corollaries and illustrated by abbreviations of the diagram at the top of page 77.

What you say is not what you saw.

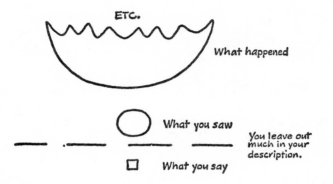

What the other person says about an event is not what happened.

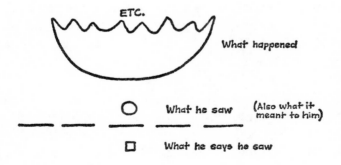

What you hear is not what the other fellow meant.

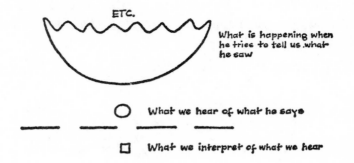

ETC.

What is happening when he tries to tell us .what he saw

What we hear of what he says

What we interpret of what we hear

In a like vein, we could state similar corollaries, such as:

What you say is not what you mean.
What you read is not what the writer wrote.
What you understand of what you read is not what the writer meant.

Consciousness of the principle of distortion opens up to you a wealth of opportunities to save time. When you feel it is important to be understood you have the choice of reducing misinterpretations by asking the other fellow what he heard of what you said. You can double-check too by asking him what he understood you to mean.

When you hear somebody say something that you can't accept, you can keep alert by asking, "Do I understand you to mean . . .?" Taking this approach with genuine feelings of interest will protect you from energy loss and sometimes give you a new and useful point of view.

When you read something you find disturbing, you can use it to further your development instead of putting it aside with annoyance. You can ask yourself first, "I wonder what the writer meant by this?" If you still find it disturbing, you can perhaps make a gain in self-knowledge by asking, "What makes me touchy on this point?" By exploring these areas of sensitivity, you sometimes may discover the source of a chronic condition of defensiveness.

When you are troubled by a problem, the principle of distortion reminds you that the problem is in actuality your description of what you perceive of what's going on. By the word "problem" you know too, that you mean you want to act but don't know the best way to act. In short, your picture suggests no acceptable line of action. When you are aware that what you describe is neither what you see nor what is going on, you have the choice of gathering more information and making new pictures until you create one that does lead to an acceptable action. A fast way of getting new pictures is open to you also when you remember that other people see things differently from the way you do. Usually they are happy to give you their pictures.

### The Principle of Projection

Working at the level of creative planning you keep alert to the principle of projection—that words influence your picture-making. When you talk you will strive to use word pictures that feature simple concrete nouns and verbs of action. This is not to say that you should avoid the use of generalizations. Not at all! The point is that the bigger the words you use, the greater care you must exercise to avoid damage. High order abstractions, our really big thinking tools, are designed to build foundations of principles and theories for the structure of your day-to-day work. Most of your talking throughout the day centers around specific problems. If you approach these problems with generalizations, it's like trying to plant a shrub by digging a hole in your front lawn with a power shovel. You get a hole but you lose most of your lawn. The bulk of the time is then taken up in repairing the damage.

When you use generalizations to sum up your conclusions, you will make sure that they are supported by plenty of examples. Whenever possible, you will use non-verbal techniques to supplement your talking and writing—techniques such as gestures, demonstrations, diagrams and pictures.

When you ask for information you will say, "What do you know about the situation?" and not, "What do you think about the situation?" You want facts, not opinions, unless you are buying them from a qualified expert. If you are going to meet someone for the first time, or visit a location new to you, you'll discourage people from giving you any but descriptive (factual) information. Above all, you will seek to train yourself to observe with both inward and outward silence. As you learn to do this, you'll begin to escape from the blinding glare of your own preconceptions. You will gain in your capacity to make sharp, accurate, fast pictures of what is going on. You will find that you have more and more time to do the things you want to do.

### The Creative Power of Words

Like all tools of power, words can wreak havoc if misused. We all know how much damage may be caused by speaking or writing impulsively, in anger, out of turn, in error, or slanderously. We are only beginning to realize that there are less well known, hidden ways in which words can slowly but inexorably destroy us. Words are like X-rays, they allow you to penetrate the past and the future. They let you take your interpretations of the past, shape a dream of the future, and make it come to pass. While giving you these creative powers, they exact their toll by destroying your present. You cannot explore the past or plan for the future without sacrificing the reality of the immediate moment. If we become too enamoured of our memories or of our visions of the future, we neglect managing time in the present. As you saw in Chapter 6, many people have already gone far in this direction. They live for a large part of every day preoccupied with problems that never materialize, past opportunities that can never be explored, and Walter Mitty-like daydreams that will never be fulfilled. All the while around them are myriads of possibilities that promise exciting adventures and rich satisfactions.

When we are alert we have the choice of using our master tools creatively while avoiding the dangers inherent in their misuse. We can be guided by the principles of complexity, distortion, and projection. We can turn away from the practice of butchery in our use of words, and aim for the skill of the surgeon.

**Chapter 14**

# Keep in contact

*Learn to make and maintain contact when you speak, and you'll gain a rich return in time.*

### Making Contact

It's 9:30 in the morning. Harold Pearson, Vice President, Production, has just cleared the morning's mail from his desk. He has asked his secretary to take incoming calls for the next hour as he must finish his work on the Naval Contract Proposal today. Now is the time to get it out of the way before some new demands on his time crop up. He starts working through the early clauses in the contract. The work runs smoothly and he begins to gain a grasp of the problem. He thinks of a better way of wording the central part of the proposal.

Abruptly Pearson becomes conscious of someone speaking to him. So far as he can he suppresses a strong feeling of irritation. Where is his secretary? The voice speaks again. "Sorry to bother you Harold, but we've run into a snag on that Matthews job." It is Stan Wright, head of Production Planning. Without waiting for Harold to reply, Stan plunges into the details of the problem. Try as he might, Harold finds it almost impossible to make sense out of Stan's story. His mind keeps drifting back to the Naval contract. He feels frustrated and tense. Suddenly he interrupts: "Wait a minute Stan, let's go over this thing again from the start, slowly." Reluctantly he pushes aside the Naval contract.

This incident, typical of the beginnings of many discussions, illustrates a major source of time losses in your day. Stan has barged in on his boss and started transmitting without first making contact. Certainly, he is in speaking contact but is not making sense to Harold. At the moment Stan appeared on the scene Harold was drawing satisfaction from the work at hand. Stan's entry brought about an interruption of that satisfaction. It immediately led to an increase of negative factors in the situation; feelings of irritation, frustration, tension, and, to a degree, hostility. As long as negative factors are high, Harold's energy—and hence his picture-making capacity—will be low. It is only because of Harold's self-control in this instance that any communication at all takes place.

### Possible Trends of an Interview

If, instead of Stan, the interruption had been some sales representative who had wandered in past the empty desk of the secretary, Harold might have cut off the fellow when his annoyance reached a break point. The time of both Harold and the salesman would have been invested to lower the capacity of each of them for the remainder of the day. The course of the interview could be diagrammed in this way:

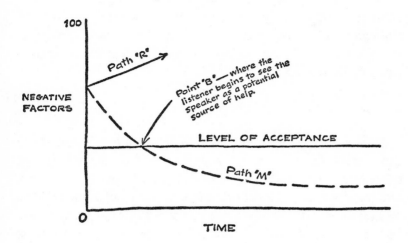

Path "R" would represent the example of the salesman's interview. Path "M" represents how an effective interview would have gone. At the start of an interview the listener can often be expected to be high in negative factors. He is being interrupted, surprised, presented with something new, or the like. (This will not be so where the interview is by appointment and the caller is expectantly awaited.) The first task of the speaker is to ensure that he has the attention and interest of his listener. This is shown as point "B" in the diagram—the point where the listener begins to see the speaker as a potential source of help. At this time the listener is probably saying to himself: "Say, this is interesting," or "I never thought of that before," or "Maybe this fellow can help me." The point "B" is on the dotted line which I call the level of acceptance. Until this level is reached there cannot be effective transmission of a picture. If you want to create time, then, you will wait until you are sure your listener is with you before you begin to transmit.

By watching your listener you can get pretty reliable clues as to his level of negative factors. By giving attention to the changing pattern of your listener's body sets, you can get a steady stream of information about his interest and understanding of your message.

### Keeping in Contact

Even after you've reached the level of acceptance something may happen to move your listener beyond your range again. A telephone call might disturb him. Something you say may start him daydreaming. Something he says may get him irritated and tense. Whatever the incident, your primary task is to get your listener back below the level of acceptance. This means first of all to stop telling your story. Second, it means to make a fast check of the Thinking, Feeling, Moving, Electro-Chemical, and Environment areas to see where a change might produce a lowering of negative factors. Perhaps there's something you can say that will calm him down. More likely, a deliberate attempt to relax yourself by opening your hands, speaking slowly, or leaning back in your chair will help him to relax too. Maybe a cup of coffee or a drink is in order; perhaps a change of scene. In any event, until you have your listener with you once again, you will only waste the time of both of you by going on talking about your idea.

By watching yourself, too, you can save a good deal of time by not allowing people to transmit to you when you cannot receive. If Harold had stopped Stan at the moment of interruption and said, "Stan, can what you have to say wait for half an hour?" he could have kept his energy high and accomplished both tasks in a fraction of the time. If the problem had to be dealt with at once, Harold might have made receiving easier by getting up from his desk and thus breaking the attention pattern he had set up for his previous task. There are some occasions during the day when you must break off what you are doing in order to listen to someone. At these times it will usually pay to give all your attention first to making sure that negative factors are relatively low. If necessary, delay the speaker until you are ready to receive.

There will be few of your listeners throughout a business day who can keep alert to all of what you are saying. If you become preoccupied, you miss the tell-tale indicators of preoccupation and negative factors in the other person. When this happens your message is lost. If you are trying to sell a product or plan, your listener is unlikely to buy on the basis of only fragments of your story. When speaking or listening, then, make sure you are in contact and you will increase your output and save time.

Chapter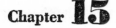

# Insure your investments of time

*Participation is a way of insuring your investments of time in new products, methods, and facilities.*

### Creating Value

When I was twelve years old my father helped me towards the purchase of a sailboat. Dad put in enough extra to allow me to buy a fifteen and a half foot racing boat of the Snipe class. I christened her the *Viking* and spent every spare moment I could with her.

After the first summer I decided that I would give the *Viking* the best finish of any boat in the fleet. I began the refinishing job as soon as she dried out in the boathouse. First came the job of burning and scraping every square inch of paint off her hull. Then came hours of sanding by hand. With the wood as smooth as I could get it, I began to go over the hull inch by inch, filling in every crack, groove, and dent with a compound of white lead and varnish, made into a thick paste. The mixture dried to the hardness of marble. It was terrible stuff to rub down with sandpaper, but it made a grand smooth finish, given enough determination, sweat, and sheets of sandpaper. Then came the paint. Lay on the paint, let it dry, rub it down. Lay on another coat of paint, let it dry, and rub it down again. Finally, the surface gleamed like a grand piano. Five months of work had built a lot of value into it for me.

Early in the summer, I was going away with my family for a weekend, and agreed to loan the *Viking* to Harry, a likeable fellow who was visiting a friend of mine. He assured me that he knew all about boats, and had sailed one of his own on the lake at his family's summer cottage. The day after I left Harry and a chum took the boat across the bay for a day's outing. They spotted a little cove that looked like an ideal spot for a swim and picnic. Coming in with a brisk, following wind, they pulled up the centerboard and drove for the gravel beach. Once aground, they pulled the boat up on the rocks away from the tide. It turned out that Harry's sailing had always been done in a rowboat with a small portable mast and sail. So far as he was concerned, small boats were always dragged up on the beach. He had not, however, had five months of his spare time tied up in the paint job on the bottom of this boat. The value of a paint job for him was only equivalent to an hour or so of his life—the time it took to slap a coat of paint on his rowboat.

The value of anything is measured by the amount of your life that you have given for it. You prize the antique furniture you made sacrifices to buy during the early years of your marriage. Your children put their feet up on it. It has no value to them because they have no life invested in it. Some friends of mine worked for ten years to establish a prize rose garden.

Then they had to move. The new owners of the house bull-dozed the garden level and paved it. It seems that their life, hence what they valued, centered around sports cars.

## The Forgotten Element in Communication

The Hamilton Company was a manufacturer of small parts for aircraft. Prior to the Second World War it had turned out a variety of stamped metal products but had quickly switched over to subcontracting for aviation components after Pearl Harbor. Despite acute problems arising out of green labor and inadequate supervision, the business mushroomed. Costs were very high but the focus was on production.

With the war over, the company found it had to lower costs drastically in order to become competitive. A firm of management consultants was called in. Four consultants studied every phase of the operation for over three months. During this time they roughed out a new order-invoice procedure, a production control system, an inventory control system, and a work simplification program. Keeping pretty much to themselves they called on managers and employees for information only when it was required. At the end of their study they prepared a large, bound report on their findings, and presented it to the Hamilton Executive Committee in a one-day session.

The presentation was masterfully planned, and the committee members unanimously agreed at the end of the meeting that the job should go ahead immediately. During the following year a team from the consulting firm worked out all the details of the new systems. Flip charts, discussion sessions, and practice runs got the inventory control system off first. The procedures were carefully explained to each group of employees as they became affected. With a consultant hovering over the system it seemed to work fine. When he left, things didn't go quite so smoothly. It wasn't that anyone was deliberately sabotaging the effort, but that a lot of people just didn't give a damn about it. They had not been given much of an opportunity to contribute to the plans so they didn't value them very highly. The system had been well designed. It had been well presented. But a key element had been left out. The forgotten element was *value*. No value had been communicated to the people who were to use and care for the system; therefore they neglected it, patched it up occasionally with makeshift changes, and eventually abandoned it. Thousands of man hours of both managers' and workers' time had been sacrificed needlessly. The company's program of innovation was set back years.

### Insuring Your Investments of Time

Value is linked with time and effort. If the consultants had asked the Hamilton employees to work with them through each stage of the design, had encouraged them to contribute their ideas, had conducted discussion sessions around each problem as it arose, they would have transmitted value. It would then have been "our system" for each employee. Even if the design had turned out to be somewhat mediocre, the employees would have made "our system" work. People don't easily abandon their "brain children." Participation is essentially the process of building value into an idea or project. The more care and support the idea needs, the more participation should go into it.

There is a relatively unexplored field for participation in selling. People tend to buy what they value. Participation is a way of creating value. Recently I was talking to an airline official who has played a big role in the purchase of a number of large aircraft of various types. He said to me, "You know, some of the aircraft manufacturers are doing a fine job of promotion. From the start of a new design they set up committees of representatives of the large carriers to study the problems of the new aircraft. These committees work out many of the details of the design they want, they plan procedures for maintenance, overhaul, flight training, ground crew training, cargo handling, and and so on down the line. By the time the first plane is off the assembly line, the carriers have a big stake in it. To hundreds of airline employees the new aircraft is 'their Super-Connie' or 'their DC7.' "

If you want people to value an idea you've got to let them help create it, or at least spend a good deal of their time working on it. Participation is more than a current trend in managing. It is the only method by which you can be sure that employees will care for and improve a growing network of complex systems and equipment. It is the only way you can insure your investments of time in new products, methods and facilities.

Chapter **16**

# Diagramming

*To get understanding and agreement in
less time, draw a diagram.*

## Non-Verbal Techniques
### of Communication

In Chapter 13 you saw some of the factors that influence a conversation. When two or more people talk together each of them brings differing past experiences to bear on his interpretation of the situation. They look at it from different positions, with different personal makeups, with differing degrees of energy depending on how they feel during their moments of observation. Then, too, they observe with differing purposes and often against the differing backgrounds created by changes in time. If they are to work together effectively they need to work from an approximately common picture. Remember that they can only act on the basis of the picture in their mind.

How then can they arrive at agreement, at an approximately common view of the "facts" and the goal? One way of diagramming their problem would be like this:

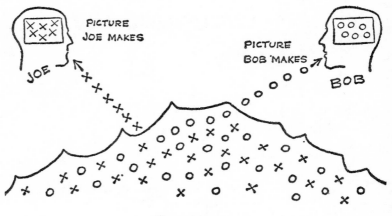

SITUATION

The fastest method of sharing a common picture is the one we discussed in the last chapter—the technique of participation. By creating an idea together, two people understand each part—they understand how the parts relate to one another, and value the final product.

The next most effective method of communication, or sharing an idea, is the technique of demonstration. By acting out the picture the listener or the viewer can imagine himself doing the same thing—he becomes a participant in imagination. The method of demonstration is widely used in business: demonstration of cars, washing machines, and vacuum cleaners. Demonstrations usually combine products or working models, gestures, (which can be thought of as spatial diagramming), and speaking. Whenever you can use a demonstration you can be reasonably sure that it will serve you well. This technique brings into play the eyes of the observer, his ears, and sometimes his senses of touch, taste, and smell. The more channels through which you transmit to him, the greater is the likelihood that your observer will understand and remember the picture you are sending.

Another range of methods for transmitting ideas is what I call the techniques of Picturing. These would include photographs, ready-made diagrams and charts, models, artists' sketches, slide projections, and motion pictures. Sometimes these are combined with sound recordings which give an added dimension to the communication. Visual and Audiovisual methods are more extensively used in business today than ever before. If properly programmed and not used for too long periods at a time, they can serve you to a much greater extent than perhaps you realize.

All the techniques so far mentioned are predominantly non-verbal. Gestures, demonstrations, pictures, and models are all much closer to what is going on than you can get with words. They are the most descriptive tools you have.

### Diagramming

There is another non-verbal technique that can save a lot of time in business. It is an approach that shares some of the characteristics of both demonstration and picturing. It is the technique of diagramming. By diagramming I mean particularly the drawing of various kinds of lines and shapes in front of your observer to represent abstract ideas. Typical of the more advanced type of diagrams is the SR Grid diagram. By advanced diagrams I refer to diagrams that have been refined through many uses, and have become powerful working tools in their own right.

### How to Start Diagramming

The late George Zipf * of Harvard used the comparison of words and symbols as tools as one of his key analogies in describing his Principle of Least Effort. In talking about tools he gave a number of examples to show that "tools find jobs to do just as jobs find tools." For example, if you keep a hammer and saw in the basement the odds are that you'll end up doing some repair work around the house sometime, and if you have a lot of calculations to perform you usually end up by buying a calculator.

If you want to develop skill in diagramming, and discover how much time it is possible to save in a day, the first step is to have diagramming tools at hand throughout the working day. Keep a pad of paper and several black and colored pencils beside you on the desk. Hang a large pad of paper on an easel in the corner of your office, or mount a chalkboard on the wall and keep various colors of chalk on the rail below it at all times. Once the tools are immediately available, somehow or other they will get involved in your discussions. You pick up the pencil and begin doodling on the paper. The other fellow grabs a pencil when he's having trouble getting an idea across to you and says, "Here, look at the problem this way." Pretty soon everybody's in the act. As one executive said to me, "You know, ever since I put that chalkboard in my office, people who come in seem to gravitate to it. Most problems are so much easier to understand when we get them into a diagram.

* G. K. Zipf, *Human Behavior and the Principle of Least Effort* (Addison-Wesley Press, 1949).

I have cut my time in meetings by half." This comment is typical of those I have heard from most managers who now use diagramming regularly. They and their subordinates swear by it as a means to reach understanding and agreement in a fraction of the time of the usual cross-the-table discussion.

### Some Hints on Methods of Diagramming

1. *Index your diagram*—Put a date location in one corner. You may want to refer to the diagram at some future time, and it's often useful to know when you did it.

2. *Use big bold strokes*—As soon as you let yourself niggle with a pencil or piece of chalk your thinking tends to get cramped.

3. *Use a soft lead pencil, a broad marking ink pen, or chalk* —These are a guarantee against niggling too.

4. *Use plenty of paper or chalkboard space*—A further way of keeping your thinking from getting narrow or static.

5. *Keep your diagrams simple*—As soon as you get too much in a diagram it ceases to help you share your ideas.

6. *Keep diagrams tidy, but not too precise*—Precision focuses too much attention on the method and takes it away from the development of the idea. Precision also tends to lead you into niggling.

7. *Keep words to a minimum*—One of the values of a diagram is that it lets you escape to some degree from the projection effect of words. Manufacture your own symbols on the spot to act as the necessary labels for your diagram.

8. *Stay away from regular geometric forms: circles, squares, triangles, rectangles*—They lead you into being too precise. Again the method tends to overshadow the ideas being diagrammed.

9. *Use a range of colors*—Colors give added dimensions for your diagrams. They also keep interest higher than do black or white lines alone.

10. *Don't correct your diagram; draw a fresh one*—Erasing takes your attention away from the development of the idea. Maybe the next sketch will suggest something else to you.

11. *Don't try to make your diagram cover everything*—Remember that no one map can represent all of any territory. Use a separate diagram to feature one group of aspects at a time.

12. *Don't take your diagrams too seriously*—Diagramming is a useful thinking and communicating tool, but remember that it is only a way of representing something. Tomorrow you'll probably think up a much better way if you're willing to let today's diagram go.

*Some Elements You Can Use*

1. *Shapes:* a) Classify by different shapes—Example: "There were two sets of factors at play in the situation."

**b)** Classify by different sized shapes—Example: "We have considered ten factors, one of which appears to be of major importance, two of secondary importance, and the remainder of only background interest.

**2.** *Lines:* **a)** Classify by line size—Example: "The controller has much more influence on the President's decisions than has the Vice-President of Sales."

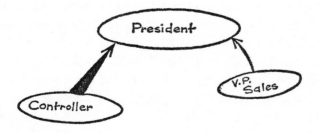

**b)** Classify by number of lines, line types—Example: "Here is a map of the chief channels of influence and key figures in the Ajax Corporation."

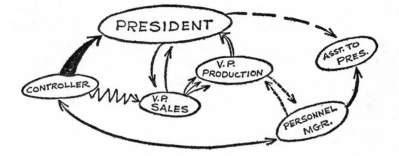

3. *Overlay:* Example: "Production and Sales share the responsibility for product planning."

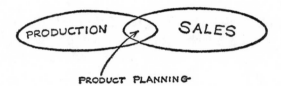

Other elements you can use are shading, cross-hatching, and colors to indicate different kinds of relationships. Arrowheads, crosses, letters of the alphabet, asterisks, numbers, and other symbols can be used in place of words to designate parts of your diagrams. Since you will be creating the diagram in front of one or more persons you can give whatever meaning you want to your signs, and this meaning will usually be remembered for the short space of time you are using the diagram. Generally speaking, diagrams are more useful for jobs that occur only once. I find that if they are used as records of a discussion, it takes too much time to write up explanatory notes on what the diagrams mean.

Once you start diagramming you will develop your own style. It is a technique that grows on you. It will give you a lot of scope for your imagination and help you discover many answers that otherwise would likely remain hidden. Adopt diagramming as a daily working tool, you'll find that managing your time in meetings will become simpler. Your listeners will say "I see" sooner.

# Chapter 17

# Compare and
# save time

*One good analogy is often worth three hours' discussion.*

### Analogies

Your little girl comes to you and says, "Daddy, what's a star?" You stop to think for a moment and then say something like this: "A star is like a big flaming ball of fire. It looks tiny because it's so far away."

The general accountant comes in with an invoice from a supplier. He says, "There's a big item here for a special die. What's a die?" You reply, "A die is like a cookie cutter or jelly mold. It lets you punch out a lot of holes in a hurry, or form metal or plastic into the same special shape any number of times."

Your little girl and the accountant were both faced with something unknown. You threw light on this unknown for them by making a comparison—a ball of fire, a cookie cutter, a jelly mold. By these comparisons, or analogies, you helped them to bridge quickly the gap to the unknown with something they already knew.

In the previous chapters I have used a number of analogies. I have spoken of your impressions as "pictures," of these pictures as "elements of experience." I spoke of negative factors as "aphids sucking away your energy," of preoccupation having taken over a continent of yourself, of the SR diagram as a "grid," of words as "containers." With each of these analogies I was trying to share with you an idea in a fraction of the time it took to create it.

Professor J. Z. Young speaks of analogies as follows:

> This whole business of making comparisons may seem to you absurd and useless. It is, however, one of our chief aids to exploring the world and hence to getting a living. Indeed, I hope to show that it is a tool we have been using in essentially the same way for thousands of years. For many purposes we have no other means of communication. It is not a question of whether or not to make comparisons but of which comparisons to make. We must use the rules—the certainties we have established by past experience. It is by comparison with these that each of us shapes his future. We *must* compare things, because that is the way our brains are constituted.*

Analogies are so fundamental in sharing your ideas that you *lean heavily* on them when you speak or write. You say, "I *grasp* the *thread* of your idea," "I *drove* home the point," "I *wore* him down," "We *strung out* the meeting," "They were *welded* into a tight group," and so on. Whenever you wish to

---

* Young, J. Z., *Doubt and Certainty in Science* (London: Oxford University Press, 1951).

*get across* an abstract idea you *borrow* many words from other fields.

There appears to be a direct relationship between your capacity to make analogies and your capacity to create new products or ideas. Your capacity to create analogies also has a big bearing on how effective you are in selling, platform speaking, or writing.

### How Analogies Can Limit Us

An analogy can be of great help in bridging the gap to something unknown, but it can also create blocks in your mind. An analogy, after all, is a comparison. It says that the unknown is like so and so. It is not, however, the thing you are talking about. Impressions are not pictures. Negative factors are not aphids. The brain is not a calculating machine. If you use an analogy too much you can easily start thinking that the analogy *is* what you are talking about. In your future observing you tend to leave out anything that doesn't fit the analogy. I read an article about one company president who fell into this trap by over-using the analogy of the "team." He became so "team-oriented" that he spent the bulk of his time recruiting a star team. In doing so he neglected product development, and his competitors left the "all-star team" far behind with a superior product. The analogy of the team has come to be a very tired symbol in business. Not only is it overworked, but it is no longer representative of modern organizational practice. Modern business is much more complicated than any game, and the team analogy fails to point up the complex inter-relatedness of business functions.

Once an analogy has served your purpose for a while, you can escape its limiting characteristics by letting go of it and creating another one to take its place. I have used the following analogy from time to time to illustrate this point: In the films of Tarzan and the apes we would see Tarzan standing on the limb of a great tree. Then he would seize a hanging vine and swing easily across to another limb, releasing the vine just before he landed. An analogy is like one of those hanging vines. It helps you swing out into the unknown but you must release it or run the risk of being swung back and smashed against the tree you just left.

### War Analogies in Business

A class of analogies which, if overused, can limit thinking in business are war analogies. Richard M. McConnell gave this example to underscore the extent to which war words are used, particularly in advertising and sales:

"Now what do we do when we begin a sales campaign? Why, we scout competition, of course. We get the lay of the land. Then we take to the field to start battering down sales resistance. We next plan a barrage of ads designed to clear away a smokescreen our rivals have laid down. Our advertising, of course, will make use of broadsides and other sales ammunition prepared by our lieutenants, our young advertising executives. Competition, knowing full well that forewarned is forearmed, have taken to a tricky defensive strategy and are busily slashing prices from coast to coast in a last-ditch stand to spearhead a drive into our sales territories. But their efforts will not prevail in the face of the bombshell we are going to unleash on our consumer target. No sir! Our sales force knows its stuff. It will get our message straight to the bull's eye every time. It is bound to be a clean sweep, a resounding victory for good old Acme Glass Tube Bending Company (not incorporated)."

Notice first of all the vintage of the war words in the above example. As McConnell points out, "they have a pre-gunpowder aroma." He says further, "They reflect a world where man's battles were more often 'hand to hand' than they are now, a world where a living had to be 'wrenched' from the soil and where death was no further away than a sword could be thrust, an arrow could fly, or a javelin be thrown."

When you use war analogies like these in your planning and sales management, what are some of the consequences? For one thing, you imply that the market is a battlefield on which one company is going to emerge the victor. It directs your attention towards beating the enemy—your competitors—and neglects the wide range of things you can do to help your customers. If you have to use war analogies in business perhaps it would be more accurate to use atomic warfare as your model. At least, it would remind you that in modern warfare everyone loses—that there are no victors, just victims.

### Fresh Analogies for Business

It seems to me that you can find many useful analogies for business in agriculture. A market can be compared to a farm land. The more carefully it is cultivated, fed, and watered, the more productive it becomes. Farmers, by exchanging their "know how," can help one another to increase their yields. If one farmer lets his fields go to seed, he creates problems for his neighbors. If he lets his crops become insect-ridden and does nothing about it, he endangers the crops of the whole region.

Think back to the principle of creating value talked about in Chapter 15. The more experience people have with something, the more they can value it. As the quantity of a good product increases in a market, people have more experience with it, and therefore value it more highly. A market grows through intensive cultivation. The steady growth of sales of life insurance is powerful testimony for this theory. Steadily the number of life underwriters licensed each year has increased. Steadily the insurance per capita in force has increased with the more intensive cultivation of this market.

Retail merchandising gives us many examples of the same kind. The more stores cluster together, the more sales tend to increase. Possibly there is a limit to the extent to which a market can be cultivated, but I wouldn't be too certain about it.

When you run into resistance to your ideas, when someone fails to see the picture you are trying to describe, when there is division in a meeting, remember the power of analogies. For example, if a manager in the sales department is balking over the new paperwork system, he might be won over with an analogy like this: "This isn't just another paperwork system, think of it as a conveyor belt that will guarantee that the customer's orders keep moving through the shop and arrive on time." In a like manner, the insurance salesman may try for a close with an analogy like this: "Picture this policy as a shield that will protect your wife and young John here, in the event that something happens to you." Whatever the block in communication you face, you can often overcome it with a good analogy. One appropriate analogy is often worth three hour's discussion.

Chapter **18**

# Read
# for growth

*What is most important is not to be able
to read rapidly, but to be able to decide
what not to read.*

### Time for Reading

Have you ever sat down and made a list of all the things you read every month, and estimated how much time you spend on each? If you do this, you'll probably find that between newspapers, magazines, and trade publications you spend at least fifty hours a month. Then you have to ask yourself this question, "Considering the some 7,000 books published in America each year, the millions of books in the Library of Congress, and the thousands of periodicals printed each month, am I getting a representative cross-section of the available information in my reading?"

Most men who ask themselves this question end up by rationing themselves to five or ten minutes a day for newspapers, ten or fifteen minutes a day for magazines of all kinds, and at least an hour a day for informative books. If they have been spending a lot of time watching TV they also ration themselves to not more than several hours a week for this medium.

### Selecting What to Read

Would you like to be able to read 50,000 words a minute? There are many times when it is easy to do this if you know how. All you have to be able to do is to recognize within one minute that a 50,000 word book does not suit your purposes, and *decide not to read it*. What is important is not to be able to read fast, but to be able to decide what not to read. *Selectivity must be the watchword.*

Here are some of the questions you can ask yourself in order to improve your selectivity:

1. What is this book about? Do I want to explore this subject now?
   —scan the table of contents and the cover flaps.

2. What are the author's qualifications? Is this the sort of person who can help me grow?
   —look at the cover flaps for biography; look at his bibliography to see how up-to-date he is, and what company he keeps.

3. How well does the author organize his material?
   —look at the table of contents, the author's preface, and a few chapter-opening paragraphs: sometimes the author has worked hard in order to make reading easy for you;

but if the book has a tortuous style, make sure there is rich pay-dirt before you start mining.

With these three groups of questions you will be able to eliminate almost all the books that would waste your time. Of course, you will miss some gems by using such a coarse screen, but you will be able to handle a big throughput and make perhaps an 80 per cent extraction.

These methods of discovering books will prove helpful:

1. Scan book reviews in high calibre periodicals, such as *Scientific American, Harpers, The Atlantic Monthly, Main Currents in Modern Thought, Science,* etc.

2. Scan the bibliographies of the books you read.

3. Browse through book stores and seek the counsel of skilled book dealers.

### Abstracting Information

You now have a book in front of you that you want to read. The first thing to remember is that no matter what you do you won't be able to get more than a few of the author's ideas. For that matter, most of the books you will read will have only a few ideas to offer that you can use at the time of reading. The trick is to find these ideas as quickly as possible. "If," as Gertrude Stein said, "a book has been a very true book for you, you will always need it again." When you come back to a book you bring to your second reading more past experience, and are able to abstract a new set of ideas.

This is a useful routine for quickly taking out of a book the information you need.

1. Read the preface and cover flaps for clues about the

author's purposes in writing the book, and the ways he uses to get his ideas across.

2. Study the table of contents and make sure you have a rough picture of the book in your mind before you start exploring it. Without this picture you can easily get lost.

3. Scan the book quickly—say in an hour or so—just in order to get to know the author and how he talks. You can't understand what a man means until you've listened to him for a while.

4. Carefully read those parts of the book that look as though they contain the information you are looking for. Often you need only read two or three chapters of a book the first time. Later, if you need more, you can go back for it. Look at a book much as you look at a hardware store; you never expect to buy up the whole store— just a few things you need at the moment.

In the appendix is a selection of some books that may make a worthwhile contribution to your growth. If you have not been doing very much reading lately, you may find it useful to limber up your "reading muscles" by first reading the books in the introductory background list. Naturally, the more books you read each month (within limits, of course), the faster progress you'll make. If you keep at it, within a year you ought to be taking a book a week into your stride easily. After that it's fair sailing and good hunting.

# SUMMARY OF PART III

1. When you can't find an answer, stop, and save time by restating the problem.
2. Talking is more than a matter of transmitting words; you speak with your whole organism.
3. Keep alert to the effects of distortion and circularity and use your language with precision and delicacy.
4. When speaking or listening, make sure you are in contact with the other person and you will increase your output and save your time.
5. Allow people to participate in the creation of your plans and ideas and they will learn to value them and care for them.
6. To get understanding and agreement in less time, draw a diagram.
7. Remember that one appropriate analogy is often worth three hours' discussion.
8. What is most important is not to be able to read rapidly, but to be able to decide what not to read.

PART **IV**

# Plan for
# Development

**Chapter 19**

# Measuring up to the big problems

*We need a broad stance in time and a workable philosophy to tackle the big problems successfully.*

### The Need for Perspective

Joe Mason was hired by Amalgamated Carbon to fill a vacancy in the service crew of the general office building. He was assigned the maintenance of the second floor of the building, and spent his first day learning the background of the job. He had to learn how the job had been done by the previous service man, where his equipment was stored, and what the people on the floor expected of him. The second day, Joe started out, full of enthusiasm, to do a bang-up job. By lunchtime he found he wasn't even a quarter of the way through his work for the day. He had been very busy but had lost a lot of time trying to find things and searching for other men on the crew who could fill him in on details of the job he thought he had understood. Within a week, however, Joe had his job in perspective. He had a feeling for all the necessary background and could service his floor comfortably in the eight-hour shift.

Like Joe, whenever we tackle a job we must get a picture of it. The bigger the job and the broader the responsibility, the greater must be the depth of our perspective. If a man comes into a top executive position from outside a company, he may spend three to six months getting a feel for the background of the organization. The larger the company, the more time he will have to spend in gaining an adequate perspective. Without this perspective he cannot, with any security, make judgments which will affect the long-range future of the company.

### The Role of the Past

In Chapter 10 we saw how the man blind from birth had to train his brain in order to see. Since he had no past experience in seeing, he couldn't form a picture of what was happening in the world by looking at it. Without a picture he couldn't act. We can only act on the basis of the pictures our past experience allows us to make. Our past experience, however, is often distorted. If we fail to re-evaluate a distorted past experience, it will prevent us from taking effective action on a current problem related to it. Here is a typical example: Ten years ago, Bob Brown, Assistant Comptroller of Selby Manufacturing, sold his boss on the idea of installing a punch card system to handle both the payroll and sales statistics. A lot of time and money was spent on installing the system.

It turned out that some of Bob's estimates of potential savings were far too optimistic. Coupled with this, it was found that the sales statistics were not being used by the sales department. The sales manager claimed that the figures he was given were not the ones he wanted (despite the fact that Bob had consulted him on his requirements), and furthermore he found they took up more time than they were worth. Bob fought for the system practically single-handedly for almost a year. Finally, he gave up and went back to a revamped version of the old system. Today, Bob is still the Assistant Comptroller of Selby. His assistant has been trying to persuade him to consider an integrated data-processing setup. Bob "knows" that punch card systems are vastly overrated. He won't even look at his assistant's plan. Bob's first unfortunate experience forms an impenetrable barrier for re-evaluating the new plan. This rigid attitude of Bob's will handicap his company under the conditions of rapid change in which it must compete. It will be a continuing source of irritation, problems, and time losses.

The role of past experience can be shown in the SR Grid in the following way:

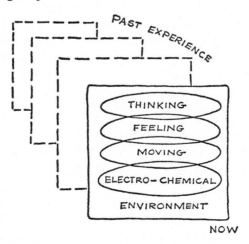

Today there is a growing realization of the importance of re-evaluating our past experience. When we or our friends suffer some emotional disturbance we more often feel free to seek the assistance of a trained counsellor or psychiatrist. Most counselling techniques are designed to help us discover and re-think over the past experiences which are bothering us in the present. In the light of our present experience we can usually put these distorted experiences back into perspective. We then regain our freedom to manage our time on occasions when our sore spots are exposed.

In a similar fashion our history is today being re-evaluated. Historians have recently awakened to the fact that most history has been based on records written almost entirely by the upper classes, and reflect only their interests. Recognizing that nine-tenths of the people were illiterate, they know too that they were voiceless. There is a huge sub-history which must be explored if we are to have a picture of the development of man rather than a history of the tiny group of priests, rulers, and their followers.

A manager's task is to plan, direct, and coordinate other people's activities. The broader his responsibility, the more people he has to take into consideration. He must deal with his employees, their representatives, his customers, suppliers, stockholders, community leaders, and politicians. He must appreciate what these people need, what they expect of him, what he can do for them, what they are likely to do next. His capacity to do this will depend to a considerable degree on how much he knows of the ways people have behaved in the past. Gordon R. Willey, Bowditch Professor of Archaeology at Harvard University, speaks of this when he says:

> The past is both inescapable and irrevocably lost. We are irresistibly swept away from it, and yet the form of the present bears the imprint of what went before. Our thoughts, our emotions, the physical shape of the world in which we live —all these are conditioned by our predecessors, and if we

are to achieve self-knowledge, we cannot disavow or disregard them. Heartening and tragic, this past is mankind's great storehouse of experience and we are wise to turn to it.

### *The Role of Our Expectations*

Not only are we guided by our past experience, but we base many of our actions on what we expect may happen. We prepare for a meeting by anticipating possible questions and objections. We buy stocks hoping they will rise. We build a house expecting to live in it, and count on our future earnings to pay for it. As we have seen in Chapter 12, we often project our fears or optimism ahead of us and discover them when we arrive at the forecast moment in time. We can complete the SR Grid by adding in dotted areas to represent the influences of our expectations on our other activities.

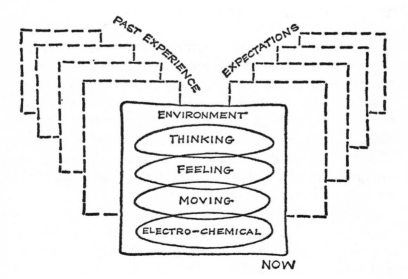

When we manage by the minute we can make spot checks on our expectations to see whether they are appropriate to our purposes of the minute. If we spend too much time dreaming about the future we are likely to fail to do now the appropriate things that would make that future come to pass. It is fairly common to find men in business, for example, so preoccupied with "getting ahead" that they're not doing their jobs. On the other hand, the higher a man climbs in an organization, the more time he must spend trying to anticipate the needs of the future. If top executives fail to do adequate long-range planning, if they are too busy to get around to it, they substitute a bumper crop of problems for the years ahead. If we walk, we need only look a few feet ahead. If we fly a supersonic jet, we must think a hundred miles in advance. Many authorities in management today are talking in terms of planning for ten, twenty, and even fifty years ahead.

### The Need for a Workable Philosophy

Our world of knowledge is now expanding at so rapid a rate that it is difficult for even a highly trained mind to comprehend the outlines of the whole. It grows not only through new scientific discoveries but through historical and archaeological research. It confronts us with more complex problems, and at the same time throws into question every belief of the past. The mood of our time is one of questioning, of uncertainty. We question authorities, morals, institutions, the value of the work we do, our purpose in life.

The great advantage of a period of turmoil is that it offers many opportunities for the development of new ideas. I watch my son playing with his Tinker Toy set. So long as it is apart and scattered about on the floor, he has the maximum set of opportunities for creating new designs. When he has all the parts fitted together into one model his opportunities for creation have been reduced to one.

Together with the creative possibilities that go with turmoil and uncertainty go great dangers as well. Only a mature person can cope with the challenge of the unexplored; he revels in extemporizing to meet the demands of the moment. The person at an earlier stage of growth becomes frightened, and hunts for a place to hide; he searches for security and is easily lured by the certainty offered by fanatical sects, extremist movements, and by the promises of big and little dictators.

There are many people in business today who are confused. They are no longer clear on the purpose of their work or their life. They feel that working for bigger production, higher sales, more take-home pay is not enough. Many of them are fearful, preoccupied with themselves, and failing to keep up with the growing demands of their work. They need more than anything else to be helped to discover a meaning for their life. Meeting this need, as I see it, is the most pressing task facing managers today. It is the task of top managers in particular to work out their own personal philosophy and then to help their people by exemplifying this philosophy in their daily behavior.

I have deliberately used the phrase "to work out" because the task is not one of finding a workable philosophy. As the South African philosopher, Rudolf Jordan, says: "The point today is that most primitive wisdom has become as insufficient and obsolete as primitive art—a museum piece—quite useless otherwise in our age. Our wisdom must be philosophical, that is, based upon the knowledge of our own generation."* If a manager goes out and begs or buys a philosophy he will have

---

* Jordan, Rudolf, *The New Perspective* (University of Chicago Press, 1951).

invested none of his life in it. Accordingly, as we saw in Chapter 15, it will have no value for him.

Creating a workable philosophy in our age of transformation is a job for strong men. When weightlifters take a big lift, they always plant their feet well apart. To measure up to the big problems a manager will need to take a broad stance in time. His mind must straddle from man's pre-history to the possibilities of the Twenty-First Century. With this preparation he can create a purpose for the managing of his time.

Chapter

# Plan to manage your time

*When you take over the direction of your personal organization you can steadily increase your available time, your output, and your satisfaction.*

### Take Control

You inherited a going concern. For the first fifteen or twenty years of your life you were heavily financed. Your resources grew steadily. Somewhere between your eighteenth and twenty-second year you were made fully responsible for your personal organization.

If today you find yourself chronically short of time you are probably under the control of your habits. Like it or not you are often led to do things that afterwards you regret. Your attention is often taken by things that really don't interest or satisfy you. You are robbed of much of your energy by a sequence of irritating incidents throughout the day.

If this is your position, you still hold the option to exercise your rights. After all, you are the sole owner of your own outfit. You can break through the control of your habits and take over as president of your own personal organization. Here is a way to do it.

### Exercise Your Rights

First you must release yourself from the feeling of being pressed for time. You can do this now and every time in the future you feel things crowding in on you again. The method suggested is crude but it works. You don't need refined techniques at this stage of the game. Follow these steps:

1. Get two slips of paper.

2. Head up one of them with the phrase "HAVE TO" and the other with "SHOULD DO."

3. List down on your "Have To" sheet *only those things you must do today.*

4. List down on your "Should Do" sheet all those things you feel you ought to do or feel it would be a good thing to do.

5. File the "Should Do" list away for future reference.

6. Take your "Have To" list and do them one by one.

The crux of this method lies in being severe in your selection of "Have To" items. Don't include anything that fails to give a "Yes" answer to this question: "Will my work, my co-workers, or my family suffer in any significant way if I fail to do this today?" You will be surprised at how few things you really have to do. You will find too, that many of your "Should Do" items will seem, on later inspection, to be quite inconsequential. If you follow this method you will have time to spare at the end of your first day. Use this time immediately to draft out some development objectives for the next three months.

### Set Quarterly Objectives

Your draft of objectives for personal development can be made in this way. Put down three headings on a piece of paper: Innovation Objectives, Growth Objectives and Maintenance Objectives. Pinpoint your specific objectives for the next three months under each of these headings in this fashion:

### Innovation Objectives

1. Select one work method to put on trial for its life. If you are a salesman, select one part of your sales presentation or routine.

2. In the coming three months subject this method to a rigorous cross-examination with these questions: (a) Can this be eliminated? If not (b) Can this be improved? And (c) Can this be delegated or done better by someone else?

### Growth Objectives

1. Decide on the number of books you will read in the next quarter and what their titles will be.

2. Decide what you will use for your daily practice session.

3. Decide which interest centering on observation you will cultivate.

4. Make a schedule of skills you will give attention to in each week of the next quarter.

*Maintenance Objectives*

1. Decide on one improvement you will make in your diet, physical activities and pattern of recreation in the next three months.

If possible, the next day show your draft of quarterly objectives to someone who is interested in your development. Enlist his or her assistance by asking for comments. Ask, as well, if you can go over your progress with him at the end of the three month period. Sharing your plans in this way will increase your likelihood of fulfilling them. Once you complete and review your first quarterly plan, immediately draft a plan for development in the next quarter.

### Check on Your Progress Daily

You wouldn't think of trying to manage a business without records. Good, simple records save your time and allow you to gauge the progress you are making. They enable you to keep a watch on the critical few elements of your activities that control the bulk of your output. Managing without records is managing without control.

The thesis of this book is that you are faced with the challenge of an increasing rate of innovation. The claim is that you can meet this challenge only if you learn to multiply your output by jumps. To do this you need to refine your techniques of self-management.

Refined techniques of management call for sensitive controls. To increase your sensitivity you will find it helpful to keep a daily log of your activities for a few months. After this time you can make a practice of reviewing in your mind your activities at the day's end.

Set aside five minutes before starting work each morning to enter your previous day's activities in your log. You can make your log in two parts: a review of yesterday's activities and your plan for the day ahead of you. Your day's plan can include a few key objectives and some reminders—reminders of routines to change, things to watch out for in your daily practice, and mental notes on where you can apply your skill of the week.

For your review of yesterday you can use the SR Grid to advantage. Your purpose is not to make a detailed report but just some notes on activities you feel were significant for your development. Here are some check questions you can use:

### Thinking Activities

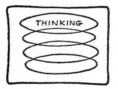

New ideas? New plans? Insights?
New interests? Gains in skills?
Reading? Reduction on talking?
Making of new diagrams? New analogies?
Listening?

### Feeling Activities

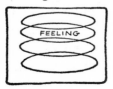

What was trend of my moods?
Attempts to overcome negative factors
Did any of my sub-personalities show up?

### Moving Activities

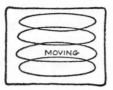

What was trend of tension during day?
What did I do?
What was tempo of my activities?
How much sleep did I get?
How did I make out in my daily practice?

### Electro-Chemical Activities

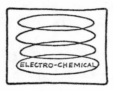

Any unusual quantities or kinds of food or drink?
Any drugs, injections, or special treatments?
Any injury, infection, sickness?
How well did I feel?

### Environment

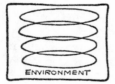

What new people did I meet?
What new places did I go to?
What new things did I experience?
What did I find distracting?
What did I do for others during the day?

### Counsel for Growth

Once you are well started on your development plan look around for someone whom you think might become interested in creating a similar plan. If you're a manager and have subordinates reporting to you, see what you can do to interest them. See that you don't coerce them into following you. You'll reap a harvest of negative factors if you do. Even though you may only be slightly ahead of them in your development, undertake to guide them *through your example.* I underscore the method of example because advice that is not asked for works equally against the receiver and the donor. Think of yourself as a mountaineer, roped to the man you are guiding. You have freely accepted the full responsibility for this man's life while on the climb. Though a freezing rain may trap you on a narrow ledge you cannot consider the possibility of giving up. He is counting on you, and your sense of responsibility to him will extend your limits of endurance. To the degree that you give of yourself in this way you will wear the cloak of invincibility. The more people you undertake to lead by example the greater will be your capacity to manage time in the face of difficulties.

Chapter **21**

## Start now

*Tomorrow, you promise yourself, will be different. Yet, tomorrow is too often a repetition of today.*

### Summing Up

The forces created by advancing technology threaten all those who refuse to develop themselves. At the same time they offer boundless opportunities to anyone who chooses to explore. You can overcome time pressures and fulfill the promise of tomorrow if you do these things:

1. Fight preoccupation by changing your routines and expanding your interests.

2. Conserve your energy by cutting down on criticism and defensiveness.

3. Expand your capacity to see by exploring the worlds of ideas and sensations.

4. Develop your skills through daily practice.

5. Support your growing resources by helping others to grow.

### The Key to Success

There is one more prerequisite to success in managing your time. The key to success in anything, as you know, is *desire*. You must have a passionate desire to develop your resources. Everything you do must become the servant of this desire. This calls to mind the reply given by the Nazarene master to the Pharisee lawyer who asked: "Teacher, which is the greatest commandment of the law?" Jesus replied: "You shall love the Lord God with all your heart, and with all your soul, and with all your mind."

Walter Russell, American philosopher, painter and sculptor, claimed that "MEDIOCRITY IS SELF-INFLICTED AND GENIUS IS SELF-BESTOWED." No matter what the level of your ability, you have more potential than you can ever fully develop in a lifetime. Regardless of your formal education, experience, position or age, it is never too late to start on the adventure of managing your time.

### Start Now

As you read these words you may not have set a time to launch your attack on habit and take over the control of your own time. If so, stop right now and ask yourself this question, "What further do I need to go ahead?" If you are honest with yourself your answer can only be, "I have all I need to start now. Furthermore, I am at this moment alert to the basis of my time problem and therefore have the power to act now. Tomorrow I may be lost in preoccupation again." Decide now to plan your development for the next three months. Now is the time to act, gain time and grow.

# Appendix

## Reading References

*Synectics,* W. J. J. Gordon, Harper & Row, New York 1961.

*The Revolution of Hope,* Erich Fromm, Bantam Books, New York 1968.

*Love and Will,* Rollo May, W. W. Norton, New York 1969.

*The Master Game,* Robert S. DeRopp, Delta Book, Dell Publishing, New York 1968.

*The Exceptional Executive,* Harry Levinson, Harvard University Press, Cambridge 1969.

*Challenge to Reason,* C. West Churchman, McGraw-Hill, New York 1969.

*The Ecology of Invasions,* Charles S. Elton, John Wiley, New York 1958.

*Beyond Economics,* Kenneth E. Boulding, University of Michigan Press, Ann Arbor 1968.

*The Ordeal of Change,* Eric Hoffer, Harper & Row, New York 1963.

*Eupsychian Management,* Abraham Maslow, Irwin-Dorsey, Homewood Illinois 1965.

*Plant in My Window,* Ross Parmenter, Apollo Edition, Thomas Crowell, New York 1962.

*Images of Hope,* William F. Lynch, SJ, Helicon Press, Baltimore, Md. 1967.

*The Age of Discontinuity,* Peter F. Drucker, Harper & Row, New York 1968.

*Education and Ecstasy,* George B. Leonard, Delacorte Press, New York 1968.

*The Bridge of Life,* Edmund W. Sinnott, Simon and Schuster, New York 1966.

*On Becoming a Person,* Carl R. Rogers, Houghton Mifflin, Boston 1961.

*The Phenomenon of Man,* Pierre Teilhard de Chardin, Collins, London 1959.

*The Nerves of Government,* Karl W. Deutsch, The Free Press, New York 1966.

*The Vital Balance,* Karl Menninger, Viking Press, New York 1963.

*The Biological Time Bomb,* G. Rattray Taylor, Thames & Hudson, London 1968.

*Stranger in a Strange Land,* Robert A. Heinlein, Berkley Medallion Book, New York 1968.

*Beyond the Outsider,* Colin Wilson, Pan Books, London 1965.

*The I Ching,* Wilhelm/Baynes, Bollingen Series, Princeton University Press, New York 1950.

*Character Analysis,* Wilhelm Reich, Farrar Straus & Giroux, New York 1949.

*The Art of Loving,* Erich Fromm, Harper & Row, New York 1956.

*Doubt and Certainty in Science,* J. Z. Young, Oxford University Press, London 1951.

*Explorations in Awareness,* J. S. Bois, Harper & Row, New York
     1957.

*Gestalt Therapy,* Perls, Hefferline & Goodman, Julian Press, New
     York 1951.

*A Study of Gurdjieff's Teaching,* Kenneth Walker, Jonathan Cape,
     London 1957.

# Index